D1165047

ALBUM OF SCIENCE

From Leonardo to Lavoisier

1450–1800

ALBUM
OF
SCIENCE

From Leonardo to Lavoisier
1450–1800

I. BERNARD COHEN

I. B. COHEN, GENERAL EDITOR, ALBUMS OF SCIENCE

CHARLES SCRIBNER'S SONS
New York

1. (*Frontispiece*) **Science and enlightenment.** In the eighteenth century, known as the Age of Reason or the Enlightenment, it was generally considered that the scientific understanding of nature and the methods of science should be part of every educated man and woman's knowledge. The illustration, taken from a general work on the arts and sciences published in London in 1759, depicts a science master or tutor instructing three young men in the principles of astronomy. He is demonstrating the planetary movements by means of an orrery, a newly invented machine that duplicates in miniature the way the planets and their satellites move in orbit. Although the scene focuses on the orrery, the stuffed animals hanging from the ceiling indicate that natural history is also important. The ship model and marine instruments on the bookcase at the right represent the study of navigation; the air pump, microscope, and telescope on the bookcase at left are for studying physics. The globe and surveyor's theodolite on the floor indicate that science has practical use, while the skeleton hanging in the closet illustrates the theme that science should be applied to the study of man himself.

Copyright © 1980 I. Bernard Cohen

Library of Congress Cataloging in Publication Data
Cohen, I. Bernard, 1914–
From Leonardo to Lavoisier, 1450–1800.
(Album of science)
1. Science—History. I. Title. II. Series.
2125.C54   509'.03   80–15542
ISBN 0–684–15377–7

1 3 5 7 9 11 13 15 17 19 Q/C 20 18 16 14 12 10 8 6 4 2

Printed in the United States of America

TO

Charles and Ray Eames

*from whom I have learned
to see the images of science
in a new perspective.*

Contents

Acknowledgments

During a two-year period while I was gathering pictures for this book, I was exceptionally fortunate in having Daniel Gale as my undergraduate research assistant. He not only kept the records and located many of the actual books or documents to be photographed, but contributed many valuable suggestions, which have been incorporated into the book. The bibliographical work was done largely by another undergraduate, Mary Susan Kish.

For aid at every stage of choosing illustrations and locating original sources I am especially indebted to William H. Bond (Librarian of the Houghton Library) and to Richard J. Wolfe (Curator of Rare Books and Manuscripts and Joseph Garland Librarian in the Countway Library of Medicine). The Russian illustrations were chosen by me from a set of pictures made available by the late Valentin Lukich Chenakal, Director of the Lomonosov Museum in Leningrad; for help in obtaining photographs for reproduction, I am particularly grateful to Achote Tigranovich Grigorian of the Institute for the History of Science and Technology in Moscow and President of the International Union of the History and Philosophy of Science. Others who have been especially helpful are H. D. Howse (National Maritime Museum, Greenwich), Eleanor Garvey (Curator of the Department of Printing and Graphic Arts in the Houghton Library), Eric Forbes (University of Edinburgh), Maurice Daumas (Conservateur-émérite, Conservatoire National des Arts et Métiers, Paris), Norman Robinson (Librarian of the Royal Society, London), David P. Wheatland (Curator of the Collection of Historical Scientific Instruments, Harvard University), John Neu (Bibliographer for the History of Science, University of Wisconsin), Lynne R. Kressly (Research Specialist, University of Pennsylvania School of Veterinary Medicine, New Bolton Center), Owen Gingerich (Smithsonian Astrophysical Observatory and Harvard University), and—above all—the Office of Charles and Ray Eames (Venice, California), especially Jehane Burns. Loren R. Graham (MIT) helped to locate and identify some of the Russian images. Georgiana D. Feldberg and Barbara J. Hughes helped check the proofs. Diana Long Hall (Boston University) helped me to select and to order the illustrations for the life sciences, and Philip Laurence gave similar help for geology.

I am particularly grateful to the photographers who have made photographic prints of books, broadsides, and manuscripts: notably Michael Nedzweski and Barry Donahue of the Fogg Art Museum (Harvard) and Steven Borack of Boston. This volume could not have been assembled without the continuing kindness of the staff of the Houghton Library (Harvard), notably the staff in the reading room headed by Martha-Eliza Shaw.

During the course of almost four decades of teaching the history of science, I have been actively engaged in searching for, identifying, and photographing (or collecting when possible) manuscripts, letters, notebooks, prints, drawings, paintings, printed books and journals, and scientific instruments in libraries, academies, private collections, museums, and print shops. I have also photographed observatories, laboratories, botanical gardens, natural history museums, scientific academies, and sites of scientific activity of the past. My aim was to give my students (and myself) a sense of the physical reality necessary for a complete comprehension of the genesis and development of the concepts we were studying.

In the course of time, I recognized that I had not merely been assembling an archive of scientific images, but had been defining an area of research on the role of illustrations in the Scientific Revolution and in the development of modern science. In the mid-1940s, I was associated with Fritz Saxl, Jean Seznec,

and Philip Hofer in exploring the potentiality and possibility of a major cooperative study in this area (with the death of Saxl in 1948, this project was stillborn). Later I had a number of opportunities to explore the relations of science and art and the nature and function of scientific illustrations with Erwin Panofsky, James Ackerman, and Samuel Y. Edgerton, Jr. (who has particularly stressed perspective for the graphic recording and dissemination of accurate scientific knowledge). Philip Hofer has, over the years, brought to my attention many examples of scientific illustration that would otherwise have escaped my attention and has made available to me the vast resources of his personal collection and the holdings of the Houghton Library (Harvard University).

For more than a decade, I also had the important opportunity to work closely with Charles Eames, who taught me by precept and example how to look at and to interpret and to make use of images in presenting scientific ideas, and with Ray Eames and the staff members of the Eames's office. I worked up my ideas concerning scientific illustrations in the Rosenbach Lectures, delivered at the University of Pennsylvania in 1973. The intellectual framework of the present book was developed as part of a research program most generously supported by the Spencer Foundation (Chicago).

This volume, and the series of which it is a part, has depended greatly on the vision and enthusiasm of Charles Scribner, Jr. I owe a great debt to a series of editors at Charles Scribner's Sons: Jane Anneken, the editor who helped launch the project; Wendy Rieder, who gave the assemblage of pictures and text their final form; and Louise Ketz, who did the preparations for the press.

Foreword

The *Album of Science* has been conceived as a pictorial record of the growth of the scientific enterprise, an attempt to show in images what science was like in the distant and recent past and to convey a sense of the perception of science by men and women, both scientists and nonscientists, living in different ages. This is the second volume in the series to appear. It deals with the establishment of modern science as we know it today, from approximately the age of Leonardo, Vesalius, and Copernicus to the time of Lavoisier, Benjamin Franklin, Volta, Linnaeus, and Albrecht von Haller. A previously published volume (by L. Pearce Williams) presents the nineteenth century. Other volumes in preparation will be concerned with ancient and medieval science, with the physical sciences in the twentieth century, and with the biological sciences in the twentieth century.

The present volume is, accordingly, not a mere record of great discoveries. That would not only be inconsistent with the general aim of the series, but it would be impossible to prepare, since some of the most important discoveries either do not lend themselves to graphic images or were never the subjects of pictorial representation. In this book, most of the major discoveries of this founding period of modern science are either the subject of an illustration or are directly related to one, or appear in the captions or the chapter introductions. Emphasis has been placed, wherever possible, on showing how scientific research and teaching were conducted and the ways in which scientific discoveries were made, rather than statements of the results of such discoveries. Thus a choice has been made in favor of pictures of laboratories and apparatus, of experiments being performed, of men and women using scientific instruments, of scientific demonstrations, of people being taught the principles of science. Since a feature of the new science was its base in experiment, pictures have been chosen to show experiments and the laboratory instruments with which experiments were performed and critical observations were made: microscopes, telescopes, chemical furnaces and retorts, electrical devices. The new vogue of natural history is represented by plans and views of botanical gardens, museums and collections of minerals and of natural history, incubators, and devices for preserving and transplanting seedlings and specimens.

No attempt has been made to include portraits of all the great scientists; those that were chosen have merits as portraits, apart from the fame of the scientist portrayed. Likenesses do appear, however, of Vesalius, Galileo, Descartes, Newton, Cassini, Linnaeus, Rumphius, Boerhaave, Buffon, and Lavoisier. Because one of the features of the new science was the emergence of scientific societies and organizations, material has been included on the ways in which scientists banded together to publish their works and to communicate their findings. As the scientific enterprise expanded, a new trade arose: the making and vending of scientific instruments, as shown by trade cards. As the sciences grew in importance, the interactions of scientists and society at large became more and more significant. This aspect of the growth of science is illustrated by scenes depicting the fate of Lavoisier and of Priestley, by the intervention of the Paris Academy of Sciences in the Mesmer affair, and by cartoons satirizing the scientist and the pseudoscientist and charlatan. One illustration shows how science was conceived to be "above the battle," an international activity that was independent of combat between nations.

In the sixteenth and seventeenth centuries, the sharp distinction had not yet been drawn between rational experimental science on the one hand, and magic, hermeticism, and alchemy on the other. This, too, is revealed in pictures of those times. But it is discernible that, early on, men of keen wit were carefully distinguishing between the hard core and the fringe and were especially concerned to draw the line between "true" science and what

was seen as the nonscientific, including astrology and some aspects of alchemy.

It will be noted that technology and invention as such are not featured in this book. The reason is a simple one: the link between the advancing front of science and a developing technology was not yet the hard and fast one that came into being during the course of the nineteenth century. The only major example of a practical invention of any significance based on scientific discovery during the period covered here was the lightning rod. Fundamental changes in the ways in which people made their living, communicated with one another, transported themselves and their goods, produced their food, cured their diseases and preserved their health, defended themselves against or attacked their enemies, manufactured items for domestic use or trade, and so on, depended on practical innovation and invention and not on the applications of new scientific discoveries. The relation between technology and science, if any, was that advances in technology helped science by providing better tools for research and, in some cases, setting problems for science, such as why suction pumps can draw up water only to a height of about 30 feet. Of course, in this period one and the same person might be both scientist and technologist; an example is Réaumur, one of the great naturalists of the eighteenth century, who wrote on steel and iron and, as a technologist rather than as a basic scientist, applied his discoveries to metallurgy. It is also true that chemists were using their knowledge in such trades as metallurgy, the making of gunpowder, and pharmacy. Christiaan Huygens and Robert Hooke applied their scientific talents to the improvement of the mechanical clock. But this does not mean that there were real breakthroughs in technological areas as a result of progress in scientific knowledge. It is for this reason that there are pictures of pumps and clocks and steam engines in this volume, but none of spinning and weaving machines or of improved methods of casting cannons. We are apt to be misled in this regard by the fact that, beginning with the founding expressions by Bacon and Descartes, scientists of the seventeenth and eighteenth centuries continued to insist that their work would yield tremendous practical fruits for industry, agriculture, and health, even though such useful applications of fundamental science were not as yet discernible.

The illustrations in this album have been chosen to indicate the great variety in available pictorial source materials. Some have been taken from books, including pages with a scientist's annotations, others from manuscripts. There are engravings, etchings, woodcuts, watercolors, pastels, paintings, drawings, and photographs of certain scientific instruments. Although the stress is on England and the Continent, there are some pictures relating to North America and South America and to the introduction of Western science into Japan and China. There are a number of images showing scientific activity in Russia, an aspect of science in the sixteenth, seventeenth, and eighteenth centuries that is all too often neglected in general histories.

The final selection of some 368 illustrations represents a choice from more than a thousand absolutely first-rate illustrations, each of which is interesting in itself and of real importance in showing what science was like in the period from Copernicus to Lavoisier and Volta. The choice in many cases has been extremely difficult. Some pictures have such great significance that they had to be included, even if well known to many readers; and yet it was also important to have fresh pictures, so as to make the book something of an adventure for all readers. I believe that there are no readers who will be familiar with all of the illustrations in this book.

I. Bernard Cohen

Part One

THE
NEW
WORLD
OF SCIENCE

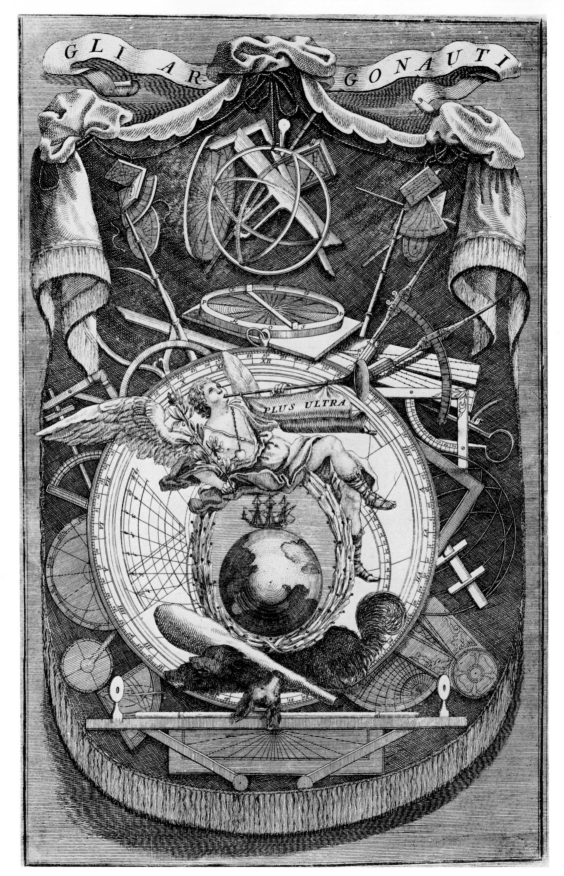

2. The boundless universe of science. The frontispiece to Marco Vincenzo Coronelli's atlas (1691) indicates that the exploration of the world depends on scientific instruments as well as on men and ships. Coronelli, a celebrated practical mathematician, mapmaker, and globemaker of the seventeenth century, fully recognized the value of such tools in navigating and mapping the earth. In contrast to the limits set on man's search for knowledge by the traditional expression *"Ne plus ultra"* ("No further"), Coronelli proclaims the spirit of the new age of science by writing *"Plus ultra"* ("Yet further") on the banner of the trumpeting angel.

1

The New World of Science

The sixteenth and seventeenth centuries were witness to the beginning of the Scientific Revolution and a point of view that held man's range of knowledge to be limitless. A familiar image in those days was the ship that carried explorers to new parts of the globe, since one of the startling novelties of the age was the discovery of new lands, peopled by human beings who were different from Europeans, and with unfamiliar plants and animals. The explorer's ship became a symbol for the new knowledge based on explorations in the world of nature.

One aspect of the new science was often expressed in relation to the phrase *"Ne plus ultra,"* supposedly engraved on the Pillars of Hercules, located at the western end of the Mediterranean and marking the traditional boundaries for voyagers by ship. This expression, meaning "No further," was taken to be characteristic of the older world of learning. But now men would set forth beyond the Pillars of Hercules of the mind, being limited only by nature itself, and by man's powers of observation and experiment.

The spirit of the new science, with its ideal of knowledge founded on direct contact with nature by means of experiment and observation, was often contrasted with the older conception of knowledge as nothing but a series of glosses on ancient learning and wisdom, with nothing essentially new to be discovered. The new science revealed many things that had never been known before, and opened up realms of inquiry and knowledge that had been closed because of the lack of instruments to probe them. Thus, the telescope and the microscope opened up the exploration of natural realms that had been hidden from man's view before the seventeenth century. But there were also new ways of looking at the world that did not depend so much on particular instruments as on adopting a new point of view, or putting on a new "thinking cap," as one historian has said. Copernicus, for example, did not base his system upon new findings obtained from the telescope or from any other instrument of observation that had not been known to his predecessors. Nor, for that matter, did he actually require observations that had not been available for more than a millennium. What his system required was a shift in point of view, which enabled him to relate celestial motions to the sun as a fixed center, rather than to the earth.

As will be seen in the chapters that follow, new ways of representing knowledge accompanied the development of the new science. The invention of perspective in drawing and painting produced a kind of authenticity in representation based on a three-dimensional realism that had not been possible earlier. In the drawings or prints by Leonardo da Vinci or Albrecht Dürer, as can be seen in the examples given in the next chapter, the new techniques conveyed a sense of reality that could on occasion even transcend the lack of direct contact with nature.

To the historian, one of the most fascinating aspects of the development of the new science was that it encompassed and made use of ancient scientific knowledge and did not necessarily cast it all out. While the errors and limitations of ancient authorities were discovered and pointed out, the claim was made in many cases that the scientists of the new age had merely made improvements on older knowledge. Eventually, while such men as Euclid and Archimedes continued to be venerated, other ancient scientists, such as Aristotle, Ptolemy, and Galen, were attacked for their errors and inadequacies, and their theories and concepts were displaced by new ones. By the seventeenth century, the difference between the ancient and the modern was obvious to all and was reflected in a literature of its own, with many works written on the theme of either "The Battle of the Books" or "The Dispute Between the Ancients and the Moderns."

CHAPTER

2

Art and Science

In the sixteenth and seventeenth centuries, works of art show the various influences of trends in science. In some cases, such as *The Assumption of the Virgin* (Ill. 13) by Galileo's friend Lodovico Cardi da Cigoli, a particular scientific discovery, like that of the surface of the moon, may be imbedded in what is obviously not at all intended to be a scientific painting. By contrast, Sebastien Le Clerc's *The Academy of Sciences and The Academy of Fine Arts* (Ill. 16) conveys a very obvious message about the relationship of those two subjects.

The contribution of art to the Scientific Revolution was based primarily on the discovery of perspective, which enabled artists to develop new modes of representing the full dimensionality of nature on a flat surface. Living creatures, buildings, landscapes, and scientific objects could be rendered in a manner that conveyed a realism lacking in earlier pictures. Various devices were created to produce drawings in perspective in a semimechanical way, chiefly by using a grid of lines at right angles to one another through which an artist viewed a scene or object. He then transferred the image to a piece of paper divided by a similar grid.

The use of perspective was especially important for the new science of anatomy. A work such as Vesalius's famous treatise on the fabric of the human body (1543) is as memorable for its magnificent illustrations as for the text, which recounts the results of the author's many dissections. Indeed, it has been argued persuasively that these illustrations may have had a greater influence than the author's words.

In order to represent the human body accurately, an artist needed to have some anatomical knowledge. It is not surprising, therefore, that artists such as Michelangelo and Leonardo made anatomical studies. For most purposes, an understanding of only the bones and surface musculature was necessary; but many an artist, once launched on a study of anatomy, was moved to learn about the various organs and the way they function.

3. **Drawing in perspective.** The art of linear perspective was discovered (or rediscovered) by the Florentine sculptor-architect Filippo Brunelleschi. Sometime during the 1420s, Brunelleschi put together a number of ideas about rendering that had been developing during the preceding centuries. Then, in the mid-1430s, another Florentine humanist, Leone Battista Alberti, described in the treatise *On Painting*, for the first time, how to draw in perspective. This system of perspective was derived from Brunelleschi, whom Alberti acknowledged, and was based on a system of grids. Alberti advised the artist to first make a "veil" of strings tied together like a grid of squares. In this way, he explained, the artist can "see any object that is round or in relief, represented on the flat surface of the veil." The artist looked at his subject through the grid, and then transferred the image square by square to a paper marked with a similar pattern. The artist in this illustration from a 1531 German treatise is applying the same principle in a simpler form, through the grid-like panes of a window.

The variety of technical problems faced by artists brought them into contact with numerous branches of science. Anatomy was associated with physiology, pigments with chemistry. The study of stresses and strains in architectural design was related to physics, and accurate renditions of the heavens drew on the new astronomy and cosmology.

Everywhere in nature, men of the sixteenth and seventeenth centuries sought mathematical relationships. Today mathematics implies a set of equations, but then it was more apt to conjure up a set of geometric figures, which could have important implications for studies of nature and the heavens. Kepler's fantasy of the relationship between the five regular solids of geometry and the orbits of the planets was made comprehensible only by perspective drawings.

Much attention was paid to the geometry of man as well, and the subject of human proportions was related to geometric figures. Inspired by remarks of the Roman architect Vitruvius, a number of artists, including Dürer and Leonardo, sought to illustrate a geometry of human proportions. The study of the motion of human beings led to an analysis of curves that are formally identical to the curves used in classical astronomy to describe the planetary orbits, or epicycles.

In the eighteenth century, many artists continued to introduce scientific subjects or themes into their works. Jean-Honoré Fragonard, known chiefly for his sentimental and romantic scenes, painted a young woman with a telescope standing in a garden; and François Boucher, most often associated with illustrations for the boudoir, turned to science for some of his inspiration. Thus, science provided subject matter for artists, while they, in turn, raised the standard of scientific interpretation. Artists also acknowledged the achievements of Descartes and the place of Newton in eighteenth-century thought.

Every realm of science was recorded by the artist. Even solar eclipses and Halley's comet were captured on paper or canvas. Symbolic title pages for works on everything from natural history to medicine, anatomy, and natural philosophy were produced, in general, by the finest artistic talents available. Plant illustration during the eighteenth century was so exquisite that it became a model for succeeding generations. In the section that follows, it will be evident that artistic renderings of scientific subjects can be viewed with the same visual delight as landscapes, domestic and historical scenes, and still lifes.

4. **From blocks to human form.** In his manual of proportion (1538), Erhard Schön used imaginary wooden blocks, presumably cubes and parallelepipeds (six-sided solids whose bases are rectangular), piled one upon the other to build human figures. The effect of perspective, the apparent foreshortening of the head and shoulders, is seen at once in the way in which any individual block seems to depart from strict rectangularity.

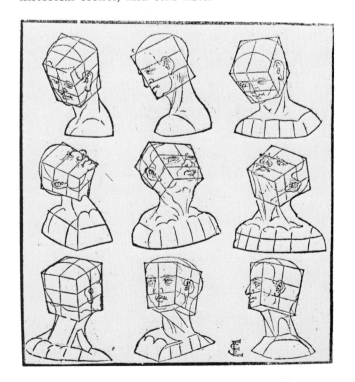

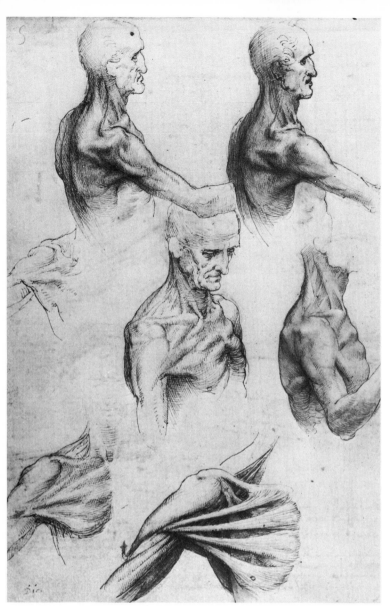

5. Art and anatomy. During the Renaissance, the influence of science on art is discernible not only in geometry, proportion, and perspective, but also in anatomy. These perspective drawings of the shoulder and arm musculature by Leonardo da Vinci show plainly why a knowledge of both anatomy and perspective was necessary to the artist who wished to make an accurate representation of the human body in different positions and performing various actions.

6. The proportions of man. A popular idea during the Renaissance was that the ideal proportions of man could be represented geometrically. The primary author of these ideas was Vitruvius, a Roman architect of the first century B.C. This illustration, taken from a 1522 edition of Vitruvius's book on architecture, is a good example of a sixteenth-century interpretation of his work.

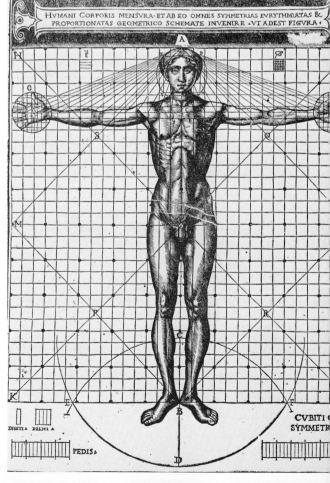

HVMANI CORPORIS MENSVRA· ET AB EO OMNES SYMMETRIAS EVRYTHMIATAS & PROPORTIONATAS GEOMETRICO SCHEMATE INVENIRE ·VT ADEST FIGVRA·

DIGITI PALMI

PEDIS

CVBITI

SYMMETR

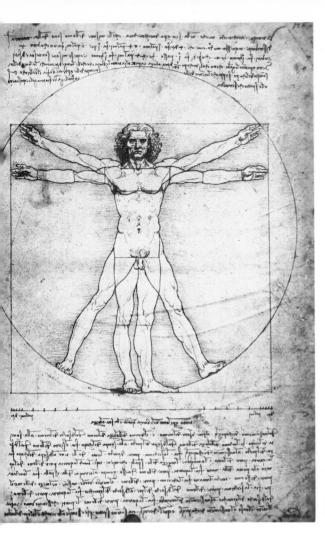

7. Leonardo's representation of the human body. In this composite diagram, Leonardo has illustrated the proportions of the human body by combining in a single illustration two of the geometric forms of Vitruvius, one based on a square and the other on a circle. To do so, Leonardo had to drop the square so that its center did not coincide with the center of the circle. Leonardo, who was left-handed, wrote backwards, that is, from right to left instead of from left to right.

8. Botticelli's St. Augustine. The Florentine painter Sandro Botticelli (1444–1510) depicted St. Augustine in his study with his bishop's miter on the desk. The striking aspect of the painting is the prominence given to three symbols of science: a mechanical clock, an armillary sphere (a model of the celestial sphere), and a geometry book. Botticelli wanted to make clear that St. Augustine was concerned with time and its measure, the nature of the celestial universe, and mathematics, as well as with his religious faith.

9. *The Ambassadors* (1533). Hans Holbein the Younger's symbolic portrait of Jean de Dinteville, who was the French ambassador to England in 1533, and his friend Georges de Selve, Bishop of Lavaur, is especially well-known for the strange shape at the bottom, which is actually a death's head. If viewed from the extreme lower left and from a point somewhat below the bottom of the picture, the skull assumes fuller shape. This use of perspective to create such an anamorphic image is a perfect example of what Shakespeare, in *Richard II*, called "perspectives, which rightly gazed upon show nothing but confusion," but when viewed "awry distinguish form." The broken lute string and the skull are reminders of death, just as the various devices for measuring time are indicative of its inexorable march. Erwin Panofsky, a leading art historian of our time, has suggested that the death's head could be further interpreted as a hidden signature, since the name of the artist, Holbein, translated literally means "hollow bone." The lute, flutes, and open hymnbook, together with a globe of the earth, a book of arithmetic, and a surveyor's square, stand for the classical combination of music and mathematics. Above them are a second globe, a sundial, and a variety of astronomical instruments. The cylindrical dial gives the date as 11 April, while the polyhedral sundial indicates the time is 10:30 a.m., a clever way of dating the painting.

10. Observation of a comet in Germany. The remarkable thing about this sixteenth-century print is that one observer of the comet, seen in 1577, is making a drawing of it while a companion holds a lantern near his tablet.

11. Dürer's fantastic rhinoceros (1515). Albrecht Dürer's magnificent woodcut of an Indian rhinoceros was based on a description and a sketch made by a Portuguese artist who had seen an actual specimen. At the time Dürer made the drawing on which the woodcut is based, he had never seen a rhinoceros, living or dead, but his artistic skill and mastery of perspective were so great that he has managed to convey a sense of authenticity in rendering this strange animal, even though this representation is characterized by fantasy. For two centuries Dürer's woodcut was copied and imitated again and again and became the standard representation of the subject in works of art and zoology. The distinctive coat of armor was the product of Dürer's artistic imagination rather than of nature itself, and has been described by the anatomist-historian F. J. Cole as "an artistic elaboration of the plicae of the skin." The small dorsal horn above the shoulder, however, is pure fantasy. The Portuguese sketch was more accurate in this detail, showing only a single horn; but Dürer and others had heard of a two-horned rhinoceros, so he may have added another horn on the withers to make sure his depiction would not be found wanting.

12. A realistic columbine. Dürer's beautiful representation of a columbine, *Aquilegia vulgaris* (1526), reveals the artist as an accurate observer of nature. It also demonstrates that art at the highest level may serve to record natural data for science and that scientific observation and the scientific principles of perspective help to enhance works of art.

13. Telescopic astronomy in religious painting. In 1612 Cigoli, a friend of Galileo and an amateur astronomer, painted this *Assumption of the Virgin* inside the dome of the papal chapel in the Church of Santa Maria Maggiore in Rome. The singular aspect of the painting is the treatment of the moon at the Virgin's feet. It is rendered as though it were seen through a telescope, in the manner Galileo recorded it just three years earlier (see Ill. 62). Thus, the light and dark portions are contrasted, so that the moon appears craggy rather than smooth, and the bright spots in the dark area, representing the reflection of the sun on mountain peaks, are included.

14. Portrait of Descartes (1649). Frans Hals's painting of René Descartes is generally considered to be a masterpiece of scientific portraiture. The imperious quality of the nose, the disdain apparent in the shape of the mouth, and the expression of the eyes agree with what we know of Descartes and denote authenticity of the highest degree. Georges Cuvier, the great French naturalist, used measurements based on the portrait to verify the authenticity of Descartes's skull, which the Swedish chemist J. J. Berzelius took to France in 1821. The skull is still preserved at the Muséum d'Histoire Naturelle in Paris.

15. "Nature Is the Book of Philosophers." The instruments pictured in this eighteenth-century engraving reflect the great range of research activity in the physical, biological, and earth sciences from the late seventeenth century to the middle of the eighteenth. A sundial at the top of the arch stands for the measurement of time; the lodestone beneath it represents magnetism. At the bottom of the picture, a collection of stones with fossil imprints symbolizes paleontology and geology, subjects to which Descartes had devoted himself. The compound microscope and magnifying glass (*lower right*), the telescope and prism (*lower left*), and the large mirror (*center left*) are developments of optical science. In the background are two tools of chemistry, a small furnace and a zigzag-shaped condensing column. The life sciences are represented by the plants, snake, crayfish, and other forms of life. Behind the mirror are a retort and an air pump, symbolizing the experimental science of matter; and at the far left stands a large machine for generating electric charges. Although natural philosophy is the theme of the picture, Descartes is still conceived as the dominant figure of science.

16. The Academy of Sciences and the Academy of Fine Arts in Paris (1698). In Sebastien Le Clerc's updated version of Raphael's *School of Athens*, the predominance of science is overwhelming. It is easy to discern the various activities of scientists connected with the recently founded Academy of Sciences: astronomy, zoology, botany, human anatomy, geodesy, hydraulics, mathematics, and theoretical and practical mechanics. A mechanical device for perspective drawing is included at the lower left.

17. *Hortus Cliffortianus.* Carolus Linnaeus's celebrated catalog of the plants in George Clifford's garden in Holland is seen at the center of this eighteenth-century painting by Jacob de Wit. The book, which is being examined by two elderly botanists and a young woman, is open to an illustration of *Turnera ulmifolia* engraved by Jan Wandelaar after a painting by George Ehret, who was one of the greatest artists of his day in the area of botanical illustration.

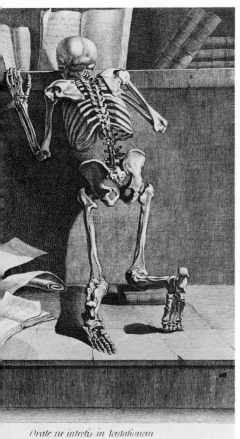

Orate ne intretis in tentationem

18. Anatomy in the service of art. Jacques Gamelin, a Toulouse artist, produced a book in 1779 that contains an extended series of plates depicting human bones and muscles. The drawings were made for the express purpose of helping artists to gain command of the body in various positions. Many of the illustrations are purely anatomical, but others, such as this one, show a skeleton assuming the pose of a living being in a realistic setting. The kneeling skeletal figure is saying in Latin "Pray that you may not enter into temptation." Gamelin's great mastery of perspective may be seen in the foreshortening of the skeletal limbs.

19. Mlle. Ferrand meditating on Newton (ca. 1753). The vogue of Newtonian philosophy in eighteenth-century France is evident in the background chosen by Maurice Quentin de la Tour for his portrait of a brilliant intellectual lady at the French court. The running head on the right-hand page of the book enables us to conclude that the book is Colin Maclaurin's *Exposition des découvertes philosophiques de Newton* (1749).

20. Boullée's project for a Newton memorial (1784). To memorialize Newton, the revolutionary French architect Etienne-Louis Boullée conceived of a gigantic cenotaph with an immense armillary sphere at the center. Boullée's studio, where he produced his architectural sketches, had portraits of both Copernicus and Newton hanging on the walls, but Newton was the inspiration for Boullée's boldest designs. In his *Essay on Art*, Boullée wrote: "Sublime spirit! Vast and profound genius! Divine being! O Newton! While you, by the scope of your insights and the sublimity of your genius, have determined the figure of the earth, I have conceived the idea of enveloping you in your own discovery."

21. The making of book illustrations. Leonhart Fuchs's *De historia stirpium* ("On the History of Plants") was published in two editions—a Latin version in 1542, followed a year later by a German translation. The text was based primarily on ancient authorities. Its importance lies in the illustrations that were carefully drawn from life. In order to stress the new aspect of this book, a special plate, shown here, was devoted to the making of the illustrations. Albrecht Meyer, who actually drew the plants from nature, is working on a drawing of a corn cockle (*upper right*), while Heinrich Füllmaurer, who transferred the drawings from paper to woodblocks, sits across from him. Below them is a portrait of Veit Rudolf Speckle, who was responsible for the actual cutting of the final woodblocks. In his preface Fuchs said: "As far as concerns the pictures themselves, each of which is positively delineated according to the features and likeness of the living plants, we have taken peculiar care that they should be as perfect as possible." Accordingly, "we have devoted the greatest diligence to make sure that every plant is depicted with its own roots, stalks, leaves, flowers, seeds, and fruits."

3

Printing

The development and dissemination of the new science was intimately related to the introduction in the fifteenth century of printing from movable type. This innovation made it possible to produce multiple copies of books, including those with illustrations, far less expensively than by hand. Before the age of printing, many scientific manuscripts had well-executed texts, but only blank spaces where illustrations were meant to be. When scientific illustrations did appear in manuscripts, they were apt to be copies of copies, or several times removed from the original and, consequently, highly inaccurate. In the case of plants, the succession of copies by illustrators who were unacquainted with the actual plants produced images in which the distinctive features of the original had been lost to such a degree that it was no longer possible to make a positive identification of the plant under discussion. With the advent of printing an illustration could be easily reproduced in multiple identical copies by a mechanical process.

In order to be printed, illustrations had to be engraved on metal plates or cut into wooden blocks. If made on a metal plate, such as copper, the illustration was printed separately and then tipped in or bound in together with the printed pages at the last stage of bookmaking. Woodcuts, however, could be printed at the same time as the text, either as part of the printed page or on the same sheet of paper, which could then be folded so that the picture appeared opposite the appropriate text.

22. A sixteenth-century printing office. Johannes Stradanus's engraving from the late 1500s shows in remarkable detail the various steps of printing. The stages from left to right are as follows. First, the copy is composed into type, which is set into a wooden case, or form. Next, a test sheet or proof is checked by the man wearing glasses, while corrections are being made to the type in a form by another man, seated beside him. Next to them, a boy is stacking the printed sheets, and behind him, a hefty man is working at the press. Note the printed sheets hanging on a line to dry. In the background, another man is inking the type.

23. Pliny's _Natural History_ (1469). Pliny's encyclopedia is devoted to the natural sciences, including geography, anthropology, botany, zoology, medicine, and mineralogy, plus certain of the arts. Printed in an edition of about one hundred copies shortly after the introduction of printing from movable type, it is considered to be the first printed book devoted to science and is one of the most beautiful books ever produced. Like many of the first books printed during the late 1400s, it was derived from a hand-copied manuscript. Many of the first scientific books to be printed were written in antiquity or the Middle Ages, not by living or recent authors. After printing, the ornamental borders and decorated capital letter "M" were drawn in by hand in every copy.

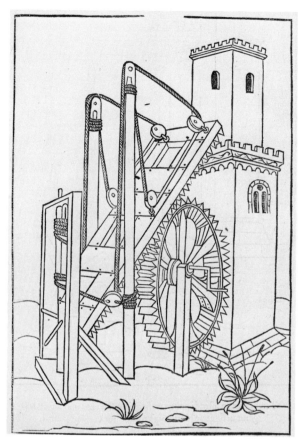

24. Military engineering. Robertus Valturius's book on military arts, *Elenchus et index rerum militarium* (probably completed between 1455 and 1460), is considered to be the earliest printed book with scientific or technical illustrations. Some of the illustrations are imaginary reconstructions of ancient instruments of warfare, but others show the equipment used in Valturius's day for assault and defense. They include cannon, portable bridges, battering rams, armed chariots, armed ships, and this rendering of a scaling ladder.

25. A world map predating Columbus. Claudius Ptolemy's *Cosmographia* is a landmark in the history of cartography. The first edition, printed in 1475, was limited to written descriptions about the earth, but the second edition, issued two years later in Bologna, was the first printed atlas as well as the first book illustrated with copper engravings. From the standpoint of printing history, however, the 1482 *Cosmographia*, published in Ulm, is the most interesting. It is the first German edition and the first to have woodcut maps. This map of Europe, Asia, and Africa comes from the 1482 edition.

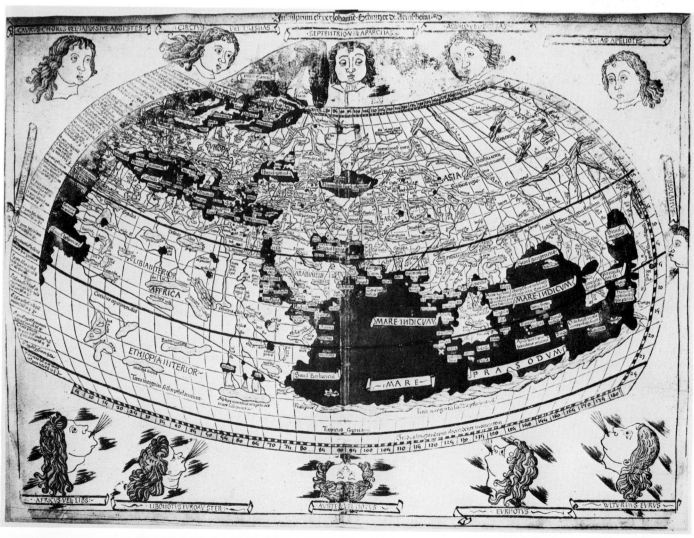

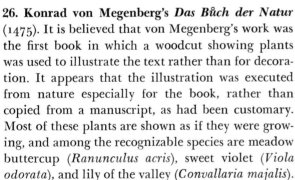

Kuchenſchell.   Hackekraut.

26. Konrad von Megenberg's *Das Bůch der Natur* (1475). It is believed that von Megenberg's work was the first book in which a woodcut showing plants was used to illustrate the text rather than for decoration. It appears that the illustration was executed from nature especially for the book, rather than copied from a manuscript, as had been customary. Most of these plants are shown as if they were growing, and among the recognizable species are meadow buttercup (*Ranunculus acris*), sweet violet (*Viola odorata*), and lily of the valley (*Convallaria majalis*).

27. Disseminating the knowledge of nature. The publication of Otto Brunfels's *Living Portraits of Plants* in 1530 is generally agreed to have inaugurated a new era in botanical illustration. Although the text is largely compiled from the work of classical and medieval authors, with some modern additions, the woodcut illustrations were derived from the careful drawings of Hans Weiditz. The plants appear as observed in nature, but without any sacrifice of beauty to accuracy. One of the great examples of Weiditz's art is this representation of the pasqueflower (*anemone pulsatilla*). Transferred to woodcuts by expert engravers, Weiditz's renditions of plants could be reproduced many times in a way that simply would have been impossible by hand. Printed illustrations were not only economical, but they also guaranteed that every copy would be as accurate and beautiful as the prototype.

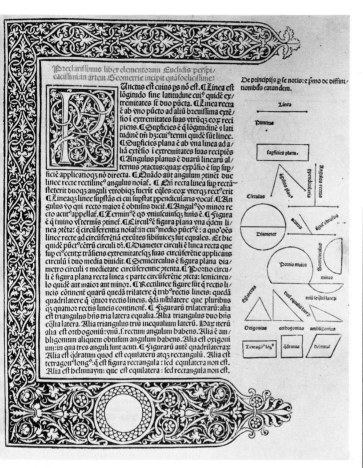

28. The first printed Euclid. In 1482 a Latin edition of Euclid's work was published in Venice. It was based on a translation of an Arabic version of the Greek text, apparently made by Adelard of Bath. The text was revised by Campanus of Novara, who added many explanations and introduced into the text some mathematical discoveries of his own era. The historian Bern Dibner has pointed out that during the first half-century of printing, there were only two editions of Euclid but seventy editions of Boethius's *Consolation of Philosophy*.

29. Archimedes in Greek. In 1544 an edition of the works of Archimedes was published in Greek and in Latin. The event was of great significance to the exact sciences of the sixteenth and seventeenth centuries. Galileo, for example, was greatly inspired by Archimedes, whose work he used and greatly praised. Note that the diagram is printed on the page with the text.

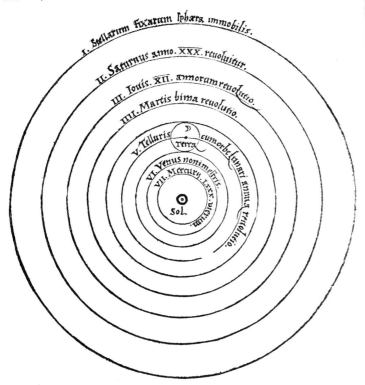

net, in quo terram cum orbe lunari tanquam epicyclo contineri diximus. Quinto loco Venus nono menſe reducitur. Sextum deniꝗ locum Mercurius tenet, octuaginta dierum ſpacio circu currens. In medio uero omnium reſidet Sol. Quis enim in hoc

I. Stellarum Fixarum Iphæra immobilis.
II. Saturnus anno. XXX. reuoluitur.
III. Iouis. XII. annorum reuolutio.
IIII. Martis bima reuolutio.
V. Telluris cum orbe lunari annua reuolutio.
VI. Venus nonimeſtris.
VII. Mercury. LXXX. dierum.
terra
Sol.

pulcherimo templo lampadem hanc in alio uel meliori loco poneret, quàm unde totum ſimul poſsit illuminare? Siquidem non inepte quidam lucernam mundi, alij mentem, alij rectorem uocant. Trimegiſtus uiſibilem Deum, Sophoclis Electra intuentē omnia. Ita profecto tanquam in ſolio re gali Sol reſidens circum agentem gubernat Aſtrorum familiam. Tellus quoꝗ minime fraudatur lunari miniſterio, ſed ut Ariſtoteles de animalibus ait, maximā Luna cū terra cognatione habet. Concipit interea à Sole terra, & impregnatur annuo partu. Inuenimus igitur ſub hac

30. The sun as center of the universe. Copernicus's *On the Revolutions of the Celestial Spheres* was published in 1543, soon after the herbals of Brunfels and Fuchs (Ills. 27 and 21). With the development of printing, scientific illustrations could be duplicated in quantity and with great clarity, and they could easily be set in the middle of a page of text as shown here. No one looking at this diagram could doubt for a moment that the doctrine expounded by Copernicus is based on the fundamental principle that the sun (*Sol*) and not the earth is situated at the center of the universe.

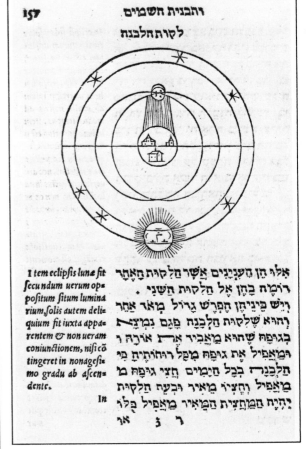

ותבנית חשמים
לסות חלבנה

Item eclipſis lunæ fit ſecundum uerum oppoſitum ſitum luminarium, ſolis autem deliquium fit iuxta apparentem & non ueram coniunctionem, niſi cō tingeret in nonageſimo gradu ab aſcendente.

אלה חן העניָנים אֲשֶׁר חֲלקות הַאתֶר
רוֹמֶה בַחֶן אֶל חֲלקות הַשֵּׁנִי ·
רֵישׁ בֵּינֵיחֶן הַחֶרְשׁ גָּרוֹל מֹאַד אַחֵר
רֵהוֹא שֶׁלקות הַלבֵנה פֵּגַם נמצָא
בְּגרֵפָה שֶׁחוֹא מַאֲבִיר אֶרֶץ אוֹרַח ר
וּמַאֲפִיל אֶת גוּפָה מֹפֵל רוּחתָרֵית בּי
תַלבֵנֵרֵ בְּכֹל חַיָמִים חֲצִי גוּפָה מ
מַאֲפִיל וַחֲצִיר מֵאִיר וּבְכֹּה חֲלקות
יֶהְיָה הַמַחֲצֵית הַמֵאִיר מַאֲפִיל פְלֹ
In אַ ר ב

31. A scientific book in Hebrew. In the fifteenth and sixteenth centuries scientific books were published in many languages: the vernaculars, Latin, Greek, and in Hebrew. A little over a century after the first printing of Euclid in Latin (Ill. 28), an edition in Arabic was printed in Rome, one of a group of remarkable books printed in Europe in Arabic letters. The page showing a lunar eclipse appeared in a treatise on astronomy in Hebrew by Abraham bar-Ḥiyya, published in Basel in 1546 by Henricus Petrus, a non-Jew. Although the text is set in Hebrew characters, there is a Latin summary in the margin.

Borderlands of Rational Science

Although the Scientific Revolution led to the Enlightenment, or the Age of Reason, it would be wrong to suppose that the development of science had cast out every vestige of superstition, irrational ideas about nature, or adherence to the ideas of charlatans. Although no astronomer or physical scientist of any reputable standing believed in astrology or alchemy by the eighteenth century, astrologers and alchemists continued to flourish and found plenty of gullible victims. It should not be forgotten that during the lifetime of Isaac Newton belief in witchcraft was still prevalent; more than one of the scientists who founded the Royal Society gave credence to witches. Common forms of prophecy and divination during the sixteenth and seventeenth centuries included palmistry, geomancy, and physiognomy, which were being practiced as late as the eighteenth century.

In considering alchemy, Hermeticism, and what was often called natural magic, it must be kept in mind that during the sixteenth and seventeenth centuries, the sharp distinction between these areas and "hard" rational experimental science was not always as obvious as it became in the eighteenth century. This does not imply that modern science is the direct descendant of mystical or irrational beliefs; yet an interest in what we now consider borderlands of rational science could, in some cases, lead to experiment and to true science. Furthermore, in the early part of the Scientific Revolution, it was still acceptable to believe that the search for truth in divine revelation or in the accumulated wisdom of ancient sages and mystics was not absolutely separated from the direct questioning of nature by experiment and observation. Natural magic aimed to unify traditional philosophy and religion with the new science of nature. For an alchemist, therefore, an oratory could be a legitimate adjunct of the laboratory, and prayer or meditation was conceived to be a necessary component of experiment.

Tycho Brahe was the last significant astronomer to be fully committed to astrology. Johannes Kepler and Galileo both cast horo-

32. Popular astrology in the eighteenth century. The country astrologer portrayed here in an Italian engraving of 1770 is about to tell a peasant woman's fortune. To demonstrate his authenticity, the astrologer carries a telescope and two books, one of which is Giambattista della Porta's treatise on physiognomy.

scopes as part of their duties as court mathematicians, although Galileo did not personally believe in the possibility of predicting a person's future from the stars. Kepler, on the other hand, was sometimes dubious about the claims of astrology, but at other times thought that astrology might provide truths about human destinies.

At a time of general belief in signs and portents, any unusual natural phenomenon could be considered a warning of imminent disaster or of a dreadful punishment to be feared unless man changed his sinful ways. Epidemics and plagues were thought to have religious significance, and diseases or malfunctions of the body were believed to be under the control of astrological signs. Even the work of the military surgeon was thought to be related to astrological conditions.

The discovery that many comets move in elliptical orbits and return at regular intervals lessened the general feeling of apprehension at their approach. Similarly, Benjamin Franklin's discovery that lightning is an electrical phenomenon and his invention of lightning rods to prevent damage transformed a frightening natural manifestation into an orderly physical phenomenon. Lightning could no longer be considered a divine warning to man.

Many of the leaders of the Scientific Revolution saw as their mission the displacement of folk beliefs in supernatural forces by scientific and rational explanations of natural phenomena. Thus Robert Recorde's *Castle of Knowledge* (1556) showed two contrasting figures—one holding the "Sphere of Destinye whose governour is knowledge" and a pair of compasses or dividers, by which all things are to be measured, and another figure blindfolded and holding the "wheele of fortune, whose ruler is Ignorance." Referring to the way in which "spitefull Fortune turns her wheele," Recorde states that "the heavens to fortune are not in thralle," and the spheres of the heavens "surmount al fortune's chance." René Descartes, in his *Discourse on Method* (1637), boasted that he had not been "taken in by the promises of an alchemist, the predictions of an astrologer, the impostures of a magician, or by the tricks and boasts of any of those who profess to know that which they do not know."

Many fantasies concerning nature survived, however, well into the Age of Reason. The existence of the jumar, supposedly a hybrid of a bull and a mare or a bull and an ass was accepted without much question by such skeptics as Voltaire and David Hume. There were also some remarkable beliefs about the life cycles of birds. It was thought that swallows hibernate under water at the bottoms of lakes or ponds. Linnaeus fully accepted this idea and wrote about it. He also believed, according to one historian, "that the rattle-snake bewitches the squirrel" and "that puppies become dwarved if they are anointed with spirits." There was also a widespread belief in those days that birds, particularly geese, could grow on trees like fruit. A number of books by naturalists of the sixteenth, seventeenth, and even the eighteenth centuries show geese being produced arboreally. Others claimed that geese come from barnacles, and this myth was first attacked seriously in 1768 by J.-É. Guettard, who pointed out that it may have originated in the appearance of the plumes, or cirri, which protrude from barnacle shells. Guettard was amazed that such an absurdity "has occupied the attention of Scientists of the first rank for centuries"; it could have been destroyed, he said, if there had ever been a "properly conducted observation" and a careful examination of the life history of barnacles. All that would have been needed would have been "to make a careful and complete dissection . . . but that would have required some care and attention and it was more convenient to give a loose rein to the imagination, especially in the case of those

Scientists who were more familiar with books than with Nature."

Alchemy, or the "hermetic art," was traditionally considered to have been founded by the lengendary Hermes Trismegistus, or "Thrice-Greatest" Hermes, in distant antiquity. The alchemists of the Renaissance drew their ideas mainly from books, but there was also an oral tradition handed down from one "adept" to another. The alchemists wrote about the transmutations of metals—processes whereby base metals such as lead could be converted into the noble metals, silver and gold. This process could be achieved, it was alleged, through a "philosopher's stone," which alchemists tried to produce. Such a stone would also cure diseases and extend the normal life span.

One of the difficulties in understanding the writings of the alchemists arises from the fact that, for the most part, they were not simply metallurgists. They often referred to metals in a metaphoric sense, in writing about the human soul or spirit and in discussing problems of religious cosmology or mystical philosophy. Recent scholars, accordingly, attempt to separate two strands of alchemy—the esoteric, dealing with spiritual, mystical, or religious matters, and the exoteric, concerned with metals. A single symbol such as *Sol*, the sun, can have both an esoteric and an exoteric meaning. Most alchemists believed that metallic processes were analogous to the living processes observed in plants and animals; they also believed in a close relationship between the great world of the universe, the macrocosm, and the smaller world of man and the earth he inhabits, the microcosm. In some cases this even led to a union between alchemy and astrology.

In the seventeenth century, alchemy took on a particular physical component, in relation to the atomic or corpuscular theory of matter. If, as such scientists as Newton and Robert Boyle tended to believe, all chemical substances or all varieties of matter are composed of the same fundamental atomic entities arranged in different ways, then it was not farfetched to suppose that chemical or alchemical reactions might reduce matter to these fundamental constituent atoms or particles, which could then be rearranged to form other varieties of matter, a process whereby one substance could be converted into another. In this sense, a transmutation of metals would be somewhat similar to ordinary chemical change. It should be observed, however, that Boyle, Newton, and John Locke, all of whom were committed to alchemical research, did not have the slightest belief in astrology.

The experiments of the alchemists led to positive knowledge in the realms we would now call chemical and metallurgical. Above all, the alchemists devised and improved much of the chemical apparatus that became standard in the seventeenth and eighteenth centuries. Some scientists of the early eighteenth century, such as Hermann Boerhaave, continued to say kind words about alchemy and alchemists, but by the second half of the eighteenth century it was hard to find a chemist—or, indeed, any other serious scientist—who was still committed to alchemy.

One of the central figures in the transition from alchemy to the modern science of chemistry and chemical medicine was Paracelsus, whose original name was Theophrastus Bombastus von Hohenheim. His attacks on Galen and the traditional medicine of the past were made with such a display of verbal pyrotechnics and ardor that his style gave rise to the adjective "bombastic," a word derived from his middle name. Paracelsus reformed chemistry and medicine, and directed alchemical thought into new directions that were fruitful both for the advance of chemistry and for the application of chemistry to medical therapeutics.

33. Casting a horoscope. In his book on divination of 1617, Robert Fludd portrayed an astrologer casting a traditional horoscope. At the right, a pair of eyeglasses lies on a table next to a book on astrology. In front of the astrologer are a celestial globe and a compass, or pair of dividers. Note that the horoscope is being cast without direct reference to astronomical or astrological tables. The world of stars governing man's fate is seen surrounding the glowing moon, even though the sun is still shining.

34. Horoscope of Johannes Schöner. The horoscope of the mathematician Johannes Schöner (born in 1477, on the 16th day of January, at 11 o'clock *ante meridiem*) was bound into a copy of Georg Joachim Rheticus's *Narratio prima* (1540). The book was addressed to Schöner, who wrote extensively on astronomy, and was the first published account of the Copernican system. Rheticus went to Poland to study with Copernicus and was primarily responsible for the publication of the latter's *On the Revolutions of the Celestial Spheres* (1543).

35. Horoscope cast for the "birth" of the Royal Observatory at Greenwich. Making fun of astrology, John Flamsteed, the first astronomer royal, cast this horoscope for the moment of "birth" of the Royal Observatory (1675 August 10^d 03^h 14^m p.m.), the time when Flamsteed himself laid the foundation stone. He added a note to the horoscope, "Risum teneatis amici," taken from Horace's *Art of Poetry*, which can be translated, "Could you refrain from laughing, my friends?"

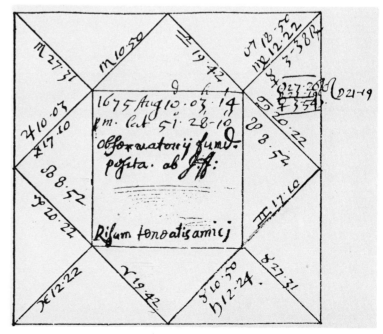

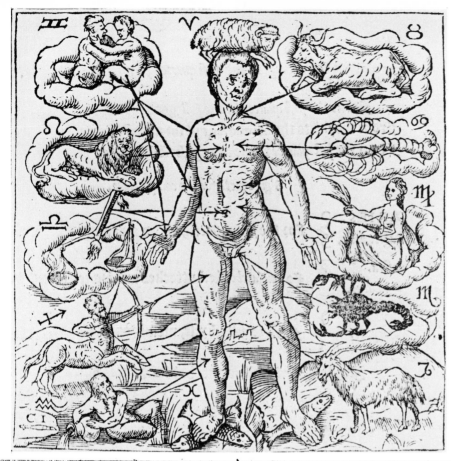

36. **Astrological medicine, from Leonard Digges's "Prognostication Everlasting"** (1556). A widespread belief held that each part of the body is controlled by one of the twelve signs of the zodiac. For example, people born under the sign of Pisces could expect to have trouble with their feet. The signs also served to indicate from which part of the body blood was to be drawn in "bloodletting."

37. **Fludd's chart of prophecy and divination** (1617). Included in Robert Fludd's book on divination is this diagram of the different arts of divination. Starting at the upper right and moving clockwise, these are: prophecy, geomancy (divination by the reading of patterns made by falling sand or earth), *ars memoria* ("art of memory," a traditional method of remembering by associating individual things with places or images; in "occult memory" this method attempts to get back to first causes), genethliacal astrology (the prediction of the future by nativities or individual horoscopes), physiognomy, chiromancy (palmistry), and *pyramidum scientia* (a method of divining based on science of the pyramids). The astrologer casting a horoscope (*bottom center*) is making an astronomical observation with a Jacob's staff; next to him is the horoscope he has cast.

38. An acephalous tribe of Indians in the New World (1599). The representation of a single acephalous man or woman might have arisen from the sight of a deformed Indian. But the representation of a whole tribe of such individuals reflects the belief that there were men and women in the New World whose heads were located in their chests rather than on necks. This fantasy was a traditional one, found in such medieval travel books as the one written by John Mandeville.

39. A jumar. In his book of 1669, Jean Léger included this illustration of a jumar and claimed: "This animal is born from a bull and a mare or of a bull and a she-ass. . . . The former have an upper jaw, nearly like pigs, but with the upper teeth an inch or two further back than those in the lower jaw; the latter, on the contrary, have a longer jaw, nearly like that of hares or rabbits." He added that "their strength is inconceivable considering their size; they are smaller than he-mules, they eat little, and they devour the distance. With such a Jumar I covered forty-five miles on the thirtieth September, entirely through mountains, and covered it much more easily than on horseback." This fictitious animal was mentioned as if it were real by such skeptics as Hume and Voltaire, and it even ended up in a later edition of Linnaeus's *Systema naturae*.

40. The grove of falsehood. In this rendering from Bartolomeo del Bene's *Civitas veri sivi morum* (1609), Falsehood is supported by a pile of windbags (or bags full of lies) and is surrounded by diviners of various sorts and a group of arrogant men who have crowned themselves with laurels. Prominent among the promulgators of falsehood are practitioners of astrology and divination and alchemists who claim they can make gold out of baser metals.

41. The laboratory and the oratory. Heinrich Khunrath included this picture of himself in his book on alchemy, published in 1609. The combination of laboratory and oratory, or chapel, indicates that the alchemist believed his science was related to the mystical union of physical processes and divine principles. The name of God appears in Hebrew characters above the oratory, and the Latin inscription on a fold of the drapery means: "When we attend strictly to our work, God himself will help us." The central table contains musical instruments and a pair of scales, symbolic of the importance of music, harmony, and number in the alchemical pursuits. The Latin inscription on the table states: "Sacred music disperses sadness [or alchemic melancholia] in evil spirits." Some chemical apparatus appears at the lower right. Although many alchemists were frauds, others sincerely explored the alchemical properties of matter and attempted to establish an analogy between the processes occurring in heaven and on earth, and between natural processes and those produced artificially by man in an alchemical laboratory. Alchemists wrote in symbolic and sometimes mystical language about the various metals and chemical substances—even endowing metals with living principles.

42. Nature, the alchemist's guide. This illustration is taken from an often re-printed collection of alchemical emblems, first published by Michael Maier in 1617. Accompanying this forty-second "emblem of the secrets of nature" was the following advice: "In pursuing chemistry, let Nature, Reason, Experiment [or experience], and Reading be the Guide, the staff, the eyeglasses, and the lamp." It is noteworthy that the chemist or alchemist, following in the footsteps of nature, wears glasses to strengthen his eyes. This is consonant with the opening words of the accompanying epigram, *Dux natura tibi* ("Let nature be your guide").

43. The "First Key" of Basil Valentine. In the symbolic representation of chemical or alchemical processes, the king or *Sol* may stand for either "sophic sulfur" or gold, and the queen or *Luna* for "sophic mercury" or silver. Here sulfur and mercury represent "esoteric" or symbolic elements, whereas gold and silver are "exoteric" or metallic. The "grey wolf" at lower left symbolizes antimony, often called by the alchemists "lupus metallorum," or the wolf of the metals, because it can "devour" or combine with all metals except gold. This property enables it to be used in purifying molten gold by "eating up" the other metals mixed with it. The old man at lower right, with a wooden leg and a scythe, represents Saturn, or lead. Beneath the wolf, gold is being purified by being fused three successive times with antimony, while silver is being "fixed" at lower right by being heated with lead. The queen holds a stalk with three flowers, symbolizing the triple purification of the king or gold. The metallic union of gold and silver, or of the king and queen, is paralleled by the combination of "sophic sulfur" (or the seed of gold) with the product of purified silver or "sophic mercury." In this way the "philosopher's stone" is to be produced, which was supposed to serve in the conversion of base metals into gold. This illustration shows the first of the twelve fundamental processes of alchemy, described in a book entitled *The Twelve Keys of Basil Valentine, the Benedictine*; the author has never been positively identified.

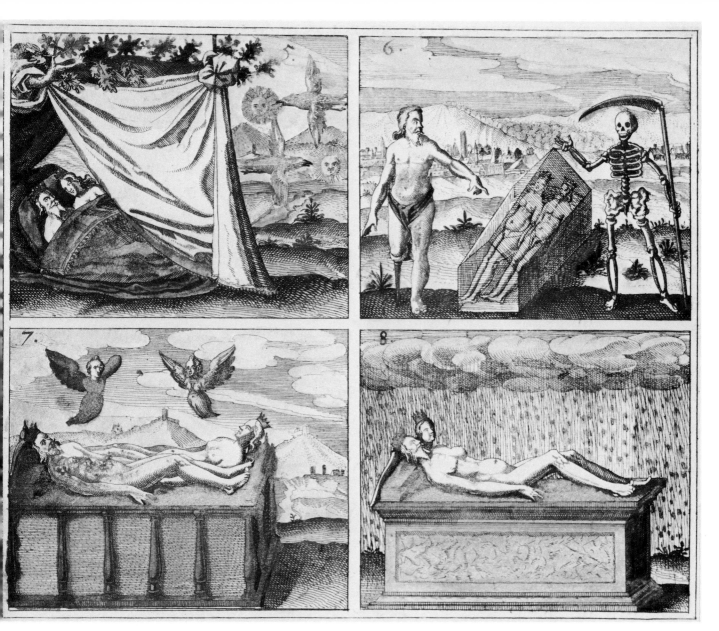

44. The alchemical death. A major symbol of alchemy is death following marriage. In this set of four pictures, the alchemical marriage (*top left*) is followed by death (*top right*). The corpses of the king and queen (*Sol* and *Luna*), as one, are enclosed within a transparent tomb. At the lower left, the bodies are becoming black and putrefied, but the grave has become a place of renewal, receptive to the celestial influence represented by birds. In the last scene the two bodies are merged into a single hermaphroditic figure. (In alchemical writings the offspring of the marriage of *Sol* and *Luna* was apt to be a hermaphrodite, often pictured in self-embrace or with arms crossed.) The ascent of mist and its descent as dew, like the rising and descending of birds, can be interpreted in terms of sublimation, distillation, and condensation—chemical counterparts of the stages by which the spirit leaves the body.

DEBENT   IGNARI   RES   FERRE   ET   POST   OPERARI   QVATVOR   INSERTA   NATVRIS   IN   NVBE   REFERTA
IVS   LAPIDIS   CARI   VILIS   SED   DENIQS   RARI   NVLLA   MINERALIS   RES   EST   VBI   PRINCIPALIS
VNICA   RES   CERTA   VILIS   SED   VBIQS   REFERTA   SED   TALIS   QVALIS   REPERITVR   VBIQS   LOCALIS

45. The alchemist. This six-teenth-century print by H. Cock was made after a drawing by Peter Brueghel the Elder. In addition to the familiar furnaces, retorts, and bellows, the picture displays the feverish activity of an alchemical laboratory. The faces of those engaged in the actual labor reflect the madness that Brueghel associated with people consumed by their search for gold. Aloof from the actual manual labors, a student of alchemy is seated on a chair before a professorial desk, while, seen through the window at upper right, his family is being received at an almshouse, the father presumably having lost his fortune and all his possessions in the vain pursuit of alchemy.

46. The alchemist as ape, or *The Pleasure of Fools*. This satirical engraving of an alchemist at work is based on a painting by David Teniers the Younger (1610–1690), who produced many pictures of serious chemists at work in laboratories. Although the alchemist's feet and tail are simian, the head more clearly resembles that of an ass.

Part Two

THE
ASTRONOMICAL
UNIVERSE

47. Armillary sphere of Antonio Santucci dalle Pomarance. Santucci's gilded model of the universe is more than 8 feet tall. It exhibits the old geocentric system, with the earth at the center surrounded by the separate spheres of the seven "planets": the moon, Mercury, Venus, the sun, Mars, Jupiter, and Saturn. These are encased by the sphere of the *primum mobile*, or first mover. Such models were used to demonstrate the nature of our universe, much as planetariums do today. This one, begun in 1588 and completed in 1593, contains the coats of arms of the Medici and the house of Lorraine, and shows how intricate and complicated the old system of the universe was thought to be.

5

The Old and New Systems of the World

At the beginning of the Scientific Revolution astronomers believed in some form of earth-centered finite universe. The sun, moon, planets, and stars were thought to have two distinct motions: first, motion in orbit around the fixed earth, and second, a participation in the daily rotation of the celestial sphere, or heavenly universe, which produces the cycle of day and night. At the beginning of the sixteenth century, the major earth-centered system of the universe was Ptolemaic, but before the century ended, two new systems, the Copernican and the Tychonic, had been introduced.

The Copernican system, formally announced on a full scale in 1543, transferred the center of the orbital motion of the planets from the earth to the sun and attributed the daily motion of the heavens to the rotation of the earth about its axis. Thus the earth had two chief motions: an annual orbiting motion around the sun and a daily rotation. Because Copernicus endowed the earth with motion, considering it to be but "another planet," his system came into conflict with the authority of Scripture. It also defied common sense, since we do not "feel" either the daily rotation of the earth or its annual motion in orbit.

Copernicus based his concept on a new way of analyzing traditional astronomical information, not on new observations. He is reported to have said that if he could make his system's predictions agree with observation to within 10 minutes of arc, he would be as happy as Pythagoras had been on discovering the famous theorem named after him. (Two stars must be about 10 minutes of arc apart to be seen as two by the average naked-eye observer.) Copernicus never realized such accuracy himself.

By the end of the sixteenth century, this 10 minutes of arc that Copernicus had conceived to be a goal of accuracy was considered to be only an approximation, thanks to the reform of the method of celestial observation by Tycho Brahe. Tycho not only introduced new instruments capable of a greater degree of exactness in positional astronomy, but he also introduced a wholly new idea into practical astronomy: the making of observations of planets every night they are visible. Up until then the positions of the planets had been determined only occasionally, in order to check a theory or to provide a value for a parameter in a theory. In particular, the very extensive series of observations of Mars made by Tycho were to prove of extraordinary value for the theoretical work of Kepler. At one time, Kepler, using Tycho's observations to test a new theory, found that he could only attain an agreement of 8 minutes of arc between his theory and the observations. This

would be far too great an error, Kepler wrote: Tycho's observations could not be wrong by such an amount.

Tycho also developed his own system of the universe. Combining some elements of traditional astronomy with certain Copernican precepts, he created a theory that was compatible with both Scripture and reasonable thought. According to Tycho, an immobile earth is circled by the moon and sun, while the planets revolve in orbit around the moving sun. Because his system bridged both the old and new worlds, it gained many adherents at first.

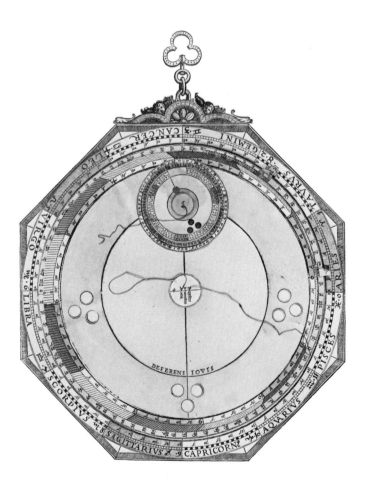

48. The complexities of the Ptolemaic system. The *Astronomicon Caesareum,* published in 1540 by Petrus Apianus, demonstrated the way in which the motion of each planet and the motion of the moon were supposed to be produced primarily by orbital motion along the circumference of a small circle (an epicycle), whose center moves along the circumference of a larger circle (the deferent). In some cases, additional circles were added as needed in an attempt to make the predictions of the system agree with observations. Each illustration in the book has a number of colored disks; in some cases there are as many as eight layers of disks that can be turned about on a thread attached to the page. The circles upon circles shown here represent the epicyclic theory of Ptolemy for the motion of the planet Jupiter. The rotation of a point in the epicycle (*top*) whose center revolves along the deferent (*Deferens Iovis*) produces a looped curve corresponding to the forward and occasionally retrograde observed orbital motion of the planet.

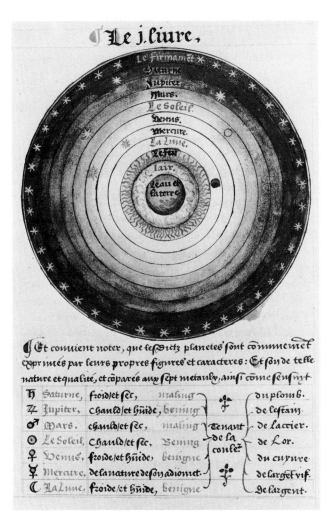

49. The traditional geocentric universe. A diagram from a holograph manuscript (1549) by Oronce Finé places the earth (*la terre*) and water (*l'eau*) at the center of the universe, surrounded by the spheres of air (*l'air*) and fire (*le feu*), representing the four Aristotelian elements. Next come the spheres of the moon, Mercury, Venus, the sun, Mars, Jupiter, and Saturn, surrounded by the firmament, or the sphere of the fixed stars. Other illustrations of the Aristotelian geocentric universe often included a "crystalline" sphere and a sphere for the prime mover outside the firmament. At the bottom of the diagram are listed the qualities and metals associated with each of the planetary spheres: Saturn, cold and dry/lead; Jupiter, hot and humid/tin; Mars, hot and dry/iron; the sun, hot and dry/gold; Venus, cold and humid/copper; Mercury, its own nature/quicksilver; and the moon, cold and humid/ silver. An indication is also given as to the benign or malignant nature of each planet.

50. Cosmography. In 1558, the beginning of the reign of Elizabeth I, William Cunningham, a doctor in physic, published *The Cosmographical Glasse*, the first work on the subject of cosmography printed in English. A diagram from the book shows Atlas supporting the universe, a combination of the spheres seen in Ill. 49, and an armillary sphere. Here, however, in addition to the spheres corresponding to the four Aristotelian elements (earth, water, air, and fire), and the seven spheres corresponding to the traditional planets, there are three external spheres for the firmament, the surrounding crystalline sphere that was thought to contain the universe, and the sphere of the prime mover.

39

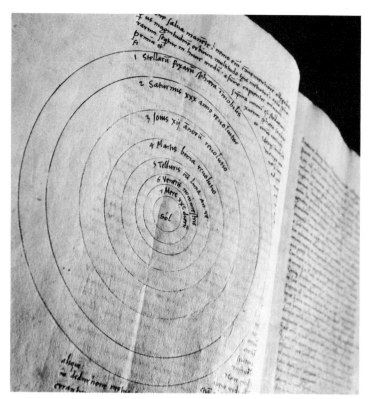

51. The essential features of the Copernican system. In his book *On the Revolutions of the Celestial Spheres* (1543), Copernicus included his own diagram of our solar system. On this page from his autograph copy, the outermost sphere (1) represents "the immobile sphere of the fixed stars." The other six spheres are for each of the six planets: Saturn (2); Jupiter (3); Mars (4); Earth, with its moon (5); Venus (6); and Mercury (7)—all circling the sun (*Sol*), which is motionless at the center. According to Copernicus, the moon moves in a circle around the earth, while the earth moves in its orbit around the sun. There is no need of a circle (or sphere) for the daily motion of the fixed stars, since, in the Copernican system, they remain at rest. The daily apparent motion of the heavens is explained by the turning of the earth on its axis. This diagram is a gross oversimplification that Copernicus introduced into the beginning of his treatise. In fact, the system is centered on the middle of the earth's orbit and not on the sun. Furthermore, Copernicus had to introduce circles moving on circles in a complexity that rivaled Ptolemaic astronomy. Kepler demonstrated the reason: planetary orbits are ellipses, not circles.

52. The finite universe. A diagram published in 1539 by Petrus Apianus shows the four central spheres of earth, water, air, and fire, surrounded by seven spheres for the traditional "planets" (which include the moon), and the spheres of the frmament, the crystalline, and the prime mover. But here there is also an indication that, just outside the sphere of the prime mover, lies the "coelum empireum, habitaculum dei et omnium electorum" ("the empyrean heaven, the habitation of God and of all the elect"). In order to account for the succession of day and night, the whole of the heavens had to rotate once in every twenty-four hours; hence, this heaven had to be finite.

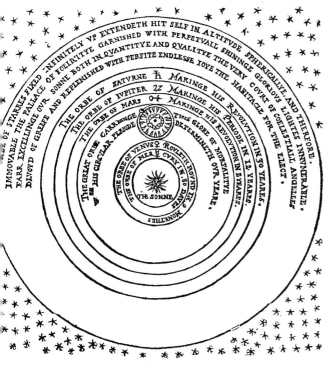

A perfit description of the Cœlestiall Orbes,
according to the most auncient doctrine of the
Pythagoreans, &c.

M 3

53. The infinite universe. A translation into English of a portion of Copernicus's book was made by Thomas Digges and first published in London in 1576. Note that the system is not called "new" and "Copernican," but rather "ancient" and "Pythagorean." Furthermore, although the "orbe" or sphere of Saturn is surrounded by the sphere of the fixed stars, which is still called the "habitacle for the elect," it is now said that this sphere is "fixed infinitely up." Copernicus himself had not said that infinite size is a feature of his system, although he had declared that the fixed stars must be at immense distances from the sun and earth.

54. Tycho Brahe, reformer of observational astronomy. Shown in his observatory on the island of Hven in 1587, Tycho is sighting a star or planet through a narrow slit in the wall at left. He is accompanied by a faithful dog. The time is determined from the clocks at the lower right and is recorded by a secretary at the table. The large brass arc cutting across the lower half of the picture was 6¾ feet in radius, 5 inches wide, and 7 inches thick. It was known as Tycho's "great mural quadrant." The two open sights attached to it were used for making celestial measurements. In the upper level of the background are various astronomical instruments used by Tycho. Below them, some students and colleagues are working at tables in the library. Some alchemical apparatus may be seen at the lowest level. Tycho's reform of astronomy was based on a long accumulation of the most accurate observations that had ever been made.

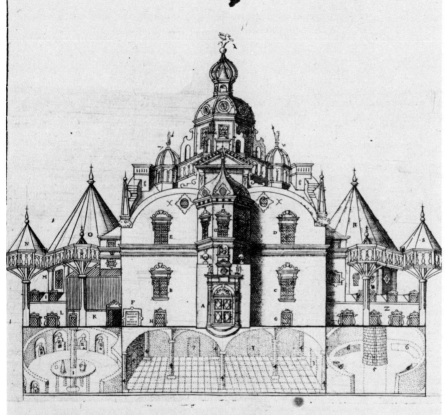

ORTHOGRAPHIA
PRÆCIPVÆ DOMVS ARCIS VRANIBVRGI
IN INSVLA PORTHMI DANICI VENVSIA *vulgo* HVENNA, ASTRONOMIÆ INSTAV-
RANDÆ GRATIA CIRCA ANNVM 1580 à TYCHONE BRAHE
EXÆDIFICATÆ.

ICHNOGRAPHIA ET EIVS EXPLICATIO

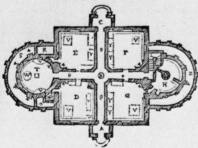

A Ianua Orientalis. C. Oc-
gulos rectos concurrentes, qui
Cœnaculum hybernum sive hypocau-
angulo post fornacem parvum quod-
gyricum esset, in quo tamen quinq,
ptius ad manus isthic operi Pyrono-
jus illud descendendum foret. B.
qui aquas hinc inde cùm lubuit, in
culum illud hybernum. E. F. G.
pro ascensu in superiorem contigna-
cementitius 40. ulnas profundus,
quas per siphones hinc inde occultè
Cameras tam superiores quam inse-
descensu in Laboratorium Chymi-
bus magnus Orichalcicus num.   exhibitus. V. Quatuor Mensæ pro Studiosis, 4. Camini tàm è laboratorio inferiori ascendentes, quam
in quatuor angulis conclavium. T. Lecti in ijsdem conclavibus, hinc inde dispositi.   Cætera acutus inspector propriâ intentione facilè
discernet. Intelligenda autem sunt hæc omnia in eâ quantitate, veluti fundamento majoris domus supra depictâ quadrare poterunt: Licet hic coarctationis loci gratiâ in duplo quasi minori formâ exhibeantur.

cidentalis. Θ. Transitus 4. ad an-
tamen posteà in tres redacti sunt, ut
stum D. ampliaretur, atq, in ejus
dam & secretum laboratorium spa-
distinctim erant furni, qui prom-
nico inserviebant, ne semper in ma-
Fons aquarium volubilem rotans,
sublime eiaculabatur. D. Cœna-
Cameræ pro hospitibus. L. Gradus
tionem. H. Coquina. K. Puteus
artificio hydraulico serviens & a-
per murum transeuntes in singulas
riores distribuens. P. Gradus pro
cum. T. Bibliotheca. VV. Glo-

55. Tycho Brahe's observatory. In 1598 Tycho published an account of his observatory, Uraniborg (castle of heaven), and of the instruments he had invented for making precise observations. Designed by Tycho, the building contained a center for astronomy and computation, but was also partly devoted to alchemical experimentation. It even had a printing press for making the results of his work available to astronomers all over the world. The four porches (marked *N*, *O*, *R*, *S* on the elevation drawing) contained instruments for making astronomical observations.

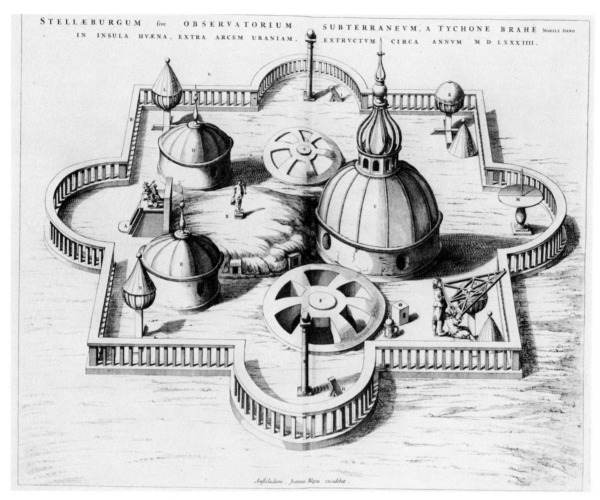

Amstelædami. Joannes Blæu excudebat.

56. Tycho Brahe's second observatory. Not far from his main observatory, Tycho designed and constructed, mainly underground, another set of buildings, which he called Stjerneborg (castle of the stars). Here important instruments could be placed "securely and firmly in order that they should not be exposed to the disturbing influences of the wind." Tycho said he also wanted "to separate my collaborators when there were several with me at the same time," so that "they should not get in the way of each other or compare their observations before I wanted this." Each of the large cupolas housed instruments for making observations of the heavens through the ground-level windows. Celestial observations could also be made through large apertures in the flat roofs of two underground chambers; these could be opened (as in *F*) or closed (as in *G*) at will. At the lower right an observer is using a huge sextant mounted on a large globe.

57. Tycho Brahe's armilla. All of Tycho's observations were made before the invention of the astronomical telescope. This particular instrument, which Tycho called "the great equatorial armilla," consisted of a declination circle, about 9 feet in diameter, and a semicircle representing the portion of the equator lying below the horizon. It was supported by eight stone piers and mounted in the largest crypt at Tycho's Stjerneborg observatory. Tycho considered the armilla to be very accurate and used the open sights to measure the positions (in altitude and azimuth) of the stars and planets.

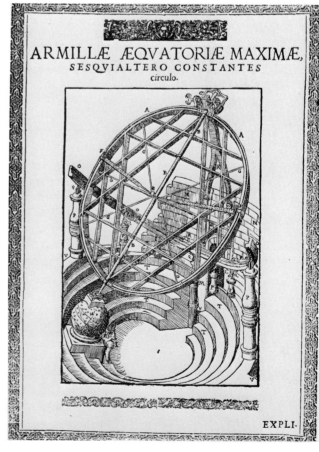

ARMILLÆ ÆQVATORIÆ MAXIMÆ,
SESQVIALTERO CONSTANTES
circulo.

EXPLI-

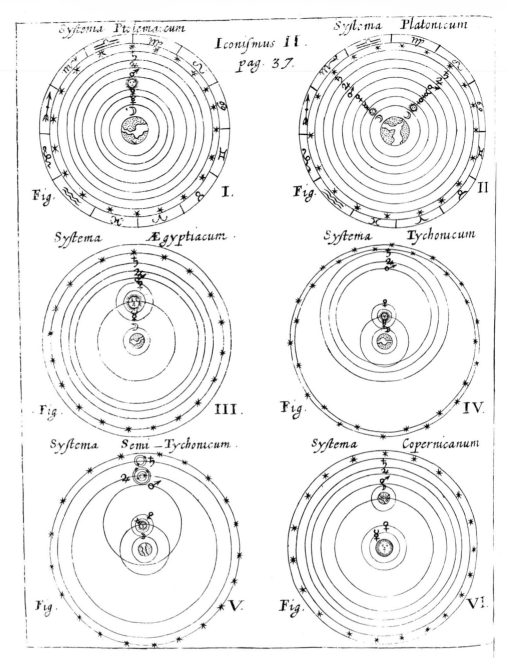

58. The six major world-systems of the late seventeenth century. In addition to
the Ptolemaic and Copernican systems (Fig. I and Fig. VI), there were four
others. According to the Tychonic system (Fig. IV), the fixed and immobile earth is
at the center of the orbits of the moon and the sun, and of the sphere of the fixed
stars (which rotates with a daily motion carrying the moon, sun, and all the planets
with it). The moving sun is the center of the orbits of the planets (the orbits
of Mercury and Venus are partly between the earth and the sun, but the orbits
of Mars, Jupiter, and Saturn encircle both the earth and the sun). The Platonic
system (Fig. II) differs from the Ptolemaic in that the two "inferior" planets, Mercury and Venus, are situated between the sun and Mars rather than between the
moon and the sun. In the so-called Egyptian system (Fig. III), the planets Mercury
and Venus revolve about the sun, while the sun, moon, and other planets revolve
about a stationary central earth. Finally, in the semi-Tychonic system (Fig. V), invented by Giovanni Battista Riccioli, three of the planets—Mercury, Venus, and
Mars—revolve about the sun as they do in the Tychonic system; but the two outermost planets—Jupiter and Saturn—have centers of revolution at the center of the
earth. Riccioli argued that Jupiter and Saturn must be distinguished from Mercury,
Venus, and Mars, since they have satellite systems like the earth, and therefore must
be anchored on the earth rather than on the sun.

59. The weighing of the systems. In 1651 Riccioli published his *New Almagest*, in which he set forth his own system of the world, a modified version of the geocentric Tychonic system. In a symbolic frontispiece, the Copernican system and a modified Tychonic system are suspended from a balance by Urania, and quite obviously the Copernican system is found wanting; it is much lighter. Underneath the balance lies the figure of Ptolemy, with his system of the universe at his feet. He is made to say, *"Erigor dum Corrigor,"* or "I am made erect by being made correct." The hand of God, at the top of the picture, indicates how the world was made by number (*Numerus*), by measure (*Mensura*), and by weight (*Pondus*).

60. Galileo, founder of telescopic astronomy. First printed as a frontispiece to his book on sunspots (1613) and later used in *The Assayer* (1623), this portrait of Galileo depicts the noted Italian scientist at the height of his powers. The cherub on the left holds the military compass, an ingenious measuring and computing device invented by Galileo. The other cherub holds a telescope, the instrument with which Galileo revolutionized astronomy. Artistic fantasy is probably responsible for the trumpet shape of the instrument. The title "Linceo" after the astronomer's name refers to the Lynxes, or the Lincean Academy, a scientific society to which Galileo belonged and that was named for its members' sharpness of eye in studying nature.

CHAPTER

6

The New Astronomical Universe

The Copernican system was difficult to accept because its fundamental postulate is that the earth is merely another planet. Most people thought the earth must be unique, different in kind and in form from the other heavenly bodies. This difference was intimately related to the singular position of the earth, motionless at the center of the universe, as set forth in the Scriptures. It was generally believed that the earth is the only body in the universe on which there is life—the only body on which God had created man.

There are also some apparent physical properties of the earth that confute the Copernican cosmology. First, the planets shine like stars, while the earth seems to be without any light of its own. Next, if the earth moved through space at the speed of about a hundred miles per minute, as postulated in the Copernican system, it would surely lose the moon. If the earth were moving, would not a ball dropped from a tower be left behind and fall to earth at some distance from the tower, contrary to all experience? Since the earth is

made of dirt, rocks, clay, and vegetative matter, surely such a body is not "fit" to participate in celestial motion! Finally, since we do not "feel" any motion, the earth must certainly stand still.

Many of these views were rudely shattered in 1610, when Galileo announced a series of spectacular discoveries made with the newly invented astronomical telescope. The telescope showed that the moon has a craggy surface and looks like a dead earth. Galileo also found that Venus exhibits phases like the moon, thus proving that the planet shines by reflected light from the sun. Correlating the phases of Venus with its apparent size under constant magnification proved that Venus must move around the sun rather than around the earth. Galileo even found evidence that the earth shines and reflects some light onto the moon. Although he never found a way to explain how the earth could move without losing its moon, he discovered that Jupiter has four moons, and he argued that if Jupiter could move without losing them, the earth could surely move without losing one moon.

Convinced of the truth of the Copernican doctrine, Galileo published the *Dialogue Concerning the Two Chief Systems of the World, Ptolemaic and Copernican* in 1632. Although he had to allege that he was merely giving the arguments on both sides of the question, he declared that "all the diseases" occur in the Ptolemaic system, while the "cures" could be found in the Copernican. For this book, Galileo was brought to trial before the Roman Inquisition, where he was condemned for having advocated and disseminated a heretical and "philosophically absurd" opinion. The book was placed on the Index of Prohibited Books, where it remained until early in the nineteenth century, when new scientific discoveries gave further support to the concept that the earth moves in orbit.

Galileo was a pioneer in formulating the view that the truths of science must be expressed in mathematics. He wrote: "Philosophy is written in that vast book which stands forever before our eyes (I mean the universe), but it cannot be read before learning the language and becoming familiar with the characters in which it is written. It is written in the language of mathematics, and the characters are triangles, circles, and other geometrical figures, without the means of which it is impossible to understand a single word."

Galileo's contemporary, Johannes Kepler, also a committed Copernican, altered the structure of astronomical theory in a number of important ways. His most radical innovation was to base astronomy on physical "causes." He held that the goal of astronomy is not merely the exact description and accurate prediction of astronomical phenomena, such as eclipses and the positions of planets. Rather, this science should reveal why the observed motions occur and should display the physical forces or causes of these motions.

Like Galileo, Kepler was a firm believer that true science must be based on a mathe-matical foundation. One of his "discoveries" was a relation between the size of the planetary orbits in the Copernican system and a nest of the five regular solids. "I undertake to prove," he wrote, "that God, in creating the universe and regulating the order of the cosmos, had in view the five regular bodies of geometry as known since the days of Pythagoras and Plato, and that He has fixed, according to those dimensions, the number of heavens, their proportions, and the relations of their movements." Kepler continued to believe in this fantasy throughout his life, but it never had any real astronomical significance.

Using the astronomical data accumulated by Tycho Brahe, whose assistant he had been, Kepler found that he could account for the apparent motions of the planet Mars only by abandoning the traditional idea that the planets move in circular orbits or orbits compounded of circles. He also found it necessary to alter one of the basic postulates of Copernicus. To simplify the mathematics, Copernicus had reckoned planetary orbits with respect to the center of the earth's orbit, rather than to the sun itself. Since Kepler insisted that all planetary motions be explained in terms of their causes, and since the sun must be the origin of the forces that move the planets around in the Copernican system, he referred all planetary orbits to the sun itself. After much labor of calculation and a great imaginative leap forward, Kepler found that planetary orbital motions can be explained to a very high degree of accuracy by the assumption that planets move about the sun in ellipses, with the sun located at one of the two foci. On the basis of his astronomical discovery of elliptical orbits, he introduced the word "focus" into the technical language of geometry. The Latin word "focus" means "hearth," and it is at a focus of an ellipse that the sun is found—at the "hearth" of the universe, so to speak.

Kepler also discovered the law of areas,

according to which the speed of a planet in orbit varies in such a way that a line drawn from the sun to the planet will sweep through the same area in the same amount of time, no matter what part of the orbit the planet may be in. This law explains or regularizes the changing orbital motion of the planets, which move quickly when near the sun and slowly when far from it. Kepler later found a third planetary law, a relation between the average distance of a planet from the sun and the time it takes to complete a revolution in its orbit. A half century later, Kepler's three laws of planetary motion provided Isaac Newton with the key to the discovery of the law of universal gravitation. These radical reforms of the Copernican concepts are so profound that what we tend to think of today as the Copernican system is really the Keplerian system.

Much time and attention were devoted in the eighteenth century to the study of comets. According to Newton's theory of universal gravitation, published in 1687, comets should move under the influence of the gravitational attraction of the sun, just as planets do. A necessary conclusion is that many comets must move in periodic orbits and return at regular intervals to the neighborhood of the sun. This concept was developed by Edmond Halley, who had been responsible for getting Newton to publish his *Principia*. Halley published a "Synopsis of the Astronomy of Comets" in the *Philosophical Transactions of the Royal Society* of London in 1705, a date that marks the beginning of modern scientific cometary astronomy.

Halley not only studied the motion of various comets, but tried to identify comets observed in the past that appeared to have returned again and again periodically. Of these, one of the most important is the comet named after him. Halley's comet is the one that appeared in A.D. 1066 and is depicted in the Bayeux tapestry. Halley said that this comet reappears at seventy-six-year intervals since it had last been seen in 1682. Halley predicted that the comet would return in 1758. When it was indeed sighted on Christmas Day 1758, astronomers hailed the event as a justification of Halley's theory of comets and Newton's theory of gravitation on which Halley based his work. In his "Synopsis" Halley had written: "Wherefore if according to what we have already said, it [the comet] should return again about the year 1758, candid posterity would not refuse to acknowledge that this was first discovered by an Englishman."

To rational thinkers of the time, the importance of Halley's prediction cannot be overestimated. One has to remember that throughout most of history, and well into the eighteenth century, comets were a source of superstitious fear on the part of many people and were considered to have been warnings from an angry God to sinful man. Halley's discovery and Newton's principles of celestial physics showed that these fearful apparitions are only simple, regularly recurring natural phenomena, which follow definable laws of nature. In addition to comets, great interest was also paid to eclipses of the sun and moon, the properties of double stars, and to meteors.

The seventeenth century witnessed the rise of the "mechanical philosophy" of René Descartes and the explanation of the celestial motions in terms of vortices of ethereal matter that swirled through space carrying the planets and other celestial bodies along with them. Although this theory continued in some form until well into the eighteenth century, its death knell was sounded in 1687 when Isaac Newton published his *Principia*, or *Mathematical Principles of Natural Philosophy*. Newton explained the motion of planets according to Kepler's laws in terms of a principle of universal gravitation, a force acting between two bodies in direct proportion to their masses and in inverse proportion to the square of the distance between them. This theory explained not only the motion of planets, but also the motion of comets, and even the action of the sun and moon to cause tides in the sea.

Although Newton envisaged using his theory to explain the irregularities of the motions of the moon, he was never fully successful in explaining and predicting all of their aspects. Computing the effects of the perturbing force of the sun on the motion of the moon required the work of many later mathematical astronomers, notably Alexis-Claude Clairaut and Leonhard Euler. The solution of the problem was a practical necessity for making accurate prediction tables of the motions of the moon that could be used in determining longitude at sea.

By the end of the eighteenth century, man's ideas of the universe began to change, owing largely to the work of William Herschel. With huge telescopes of his own design and manufacture, Herschel studied the heavens to determine the dimensions and shape of our galaxy. As a result, people began to realize that space is filled with a number of individual universes or galaxies, and that a trace of our own can be seen in the Milky Way. It never occurred to Herschel, however, that our solar system might not be located at the center of our galaxy.

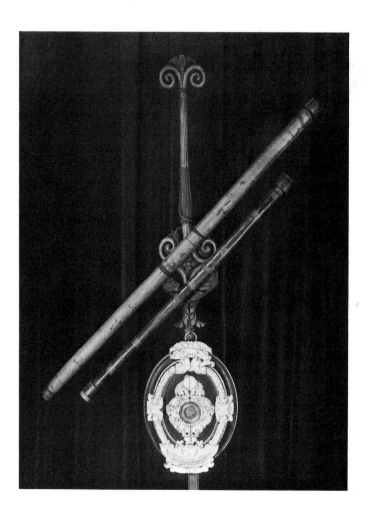

61. Galileo's original telescopes. Galileo's telescopes are now mounted on a special stand, together with one of his objective lenses, which is framed in ivory. They are preserved in the Museum of the History of Science in Florence, Italy. The telescopes consisted of a biconvex objective lens and a biconcave ocular or eyepiece, as in today's opera glasses, and were of about fourteen to twenty magnification. (An ordinary pair of today's binoculars, marked 7x50 or 8x35, has a magnification of either seven or eight.)

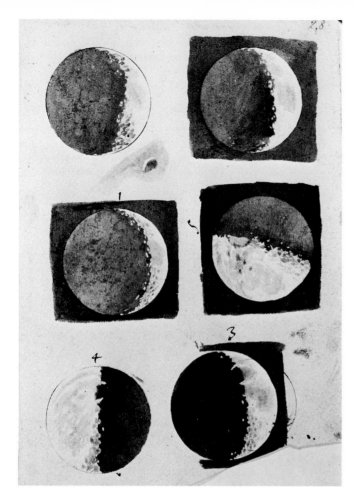

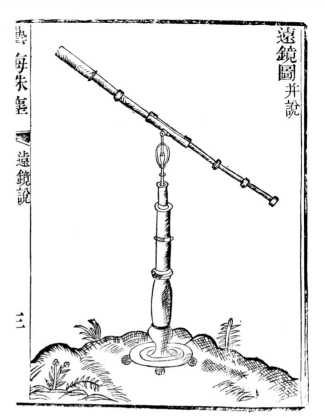

62. Original drawings of Galileo's moon observations.
Through the telescope, which Galileo improved and used
for astronomical purposes in 1609, the moon appeared
to be another earth. He discovered that the earth is not
unique: the moon has an earth-like surface, and Jupiter
also has moons. He also found that the fixed stars, unlike
the planets, do not appear as disks when viewed through
the telescope, and so concluded that they must be very far
away, just as Copernicus had taught. Galileo drew special attention to the bright spots that show up in the
dark part of the moon, explaining that they are high
mountain peaks rising out of the shadow and illuminated
by the sun. By simple geometry he correctly computed the
heights of lunar mountains to be about four miles. These
wash drawings accompany the first draft, in Galileo's
autograph, of Galileo's *Sidereal Messenger*.

64. The earliest known map of the moon. Thomas Harriot, a friend of Sir Walter Raleigh's, drew this map from
telescopic observations made in 1609 and 1610. Harriot
never published his astronomical discoveries, but was
known to his contemporaries for his work in mathematics
and for his *Briefe and True Report of the New Found
Land of Virginia* (1588). The latter was based on a visit
to the territory, which he made at Raleigh's request.

63. The telescope in China. In 1626 Johann Adam
Schall von Bell, a Jesuit missionary, produced the
first Chinese treatise on the telescope and its revela-
tions, from which this illustration is taken. Eleven
years earlier, von Bell had witnessed a convocation
in Galileo's honor in Rome. Although a Jesuit
missionary had already summarized Galileo's discoveries in Chinese in 1615, the first telescope did
not actually arrive in that country until 1618,
brought by another Jesuit, Johannes Terrentius (or
Schreck, as he was known in Chinese). The telescope was presented to the emperor of China in
1634. Jesuit missionaries found that the one area
in which they could easily convince the Chinese
of Western superiority was astronomy, particularly
with the accuracy of the Gregorian calendar.

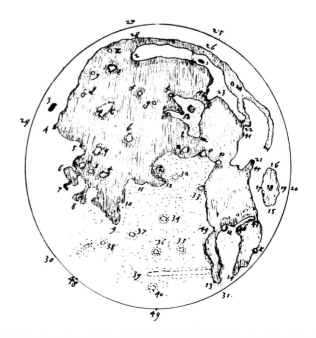

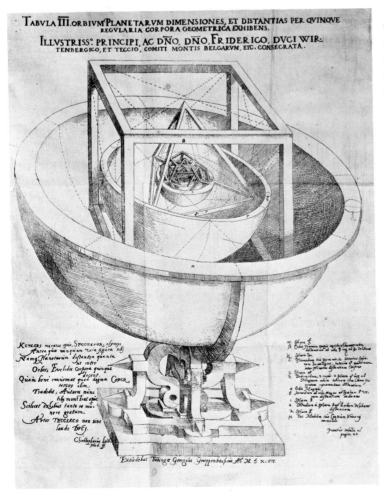

65. Galileo's observations of Jupiter's moons. In his original record of observations, on which his *Sidereal Messenger* was based, Galileo recorded the appearance of bright objects seen near Jupiter, which he correctly interpreted to be satellites or moons. They are indicated by the small star shapes alongside a star within a circle (standing for Jupiter) in several places on this page. This discovery had a two-fold consequence. Not only did it show that a planet other than the earth had a moon, but it thereby refuted the arguments of those who believed the earth could not move without losing its own moon. Galileo also found that the planet Venus exhibits phases, appearing fuller and smaller when farthest away from the earth and catching more of the sun's rays. By correlating the relative size of these phases with the position of Venus, he proved that the planet must move around the sun rather than around the earth—just as Copernicus (and Tycho) had said. Since both the earth and Venus shine by reflected sunlight and Jupiter also has moons, it appeared that the earth was not unique among the heavenly bodies, but was like the planets.

66. Geometry in the heavens. A diagram taken from Johannes Kepler's first published book, *The Cosmographical Enigma* (1596), illustrates an argument that Kepler advanced in favor of the Copernican system. It is based on the fact that there are only five regular solids: the tetrahedron (a four-sided polyhedron or pyramid whose faces are equilateral triangles), the hexahedron (six-sided polyhedron or cube whose sides are squares), octahedron (eight-sided solid whose sides are equilateral triangles), dodecahedron (twelve-sided solid whose sides are regular hexagons), and icosahedron (twenty-sided polyhedron whose sides are equilateral triangles). Kepler found that if a nest were made of the five regular solids, each pair being separated from its neighbor by a sphere, then the relative radii of the spheres would correspond to the average planetary distances according to Copernicus. In the diagram, the outer sphere for Saturn's orbit circumscribes a cube that circumscribes a tetrahedron that circumscribes the sphere for Mars's orbit. The dodecahedron is similarly related to the sphere for the earth's orbit, the icosahedron to the sphere for Venus's orbit, and the octahedron to the sphere for Mercury's orbit. Kepler's lifelong belief in this celestial geometry is of interest primarily as an illustration of his personal scientific outlook.

67. The elliptical orbits of planets. Kepler demonstrated in his *New Astronomy* (1609) that planets move in elliptical orbits with the sun at a focus. On this page from the book, he developed certain properties of the ellipse necessary for his demonstration. Proposition II on the page states that if an ellipse is inscribed in a circle, as in the diagram, then the area of the ellipse bears the same proportion to the area of the circle as the semi-axis minor of the ellipse (line *bh*) does to the radius of the circle (line *eh*). The development of the geometry of the ellipse was a major feature of the *New Astronomy*, and in it Kepler introduced the word "focus" into the geometry of conic sections. It was in this work that Kepler announced the law of areas.

289 DE MOTIB. STELLÆ MARTIS

PROTHEOREMATA.

I.

S I intra circulum defcribatur ellipfis, tangens verticibus circulum, in punctis oppofitis; & per centrum & puncta contactuum ducatur diameter ; deinde a punctis aliis circumferentiæ circuli ducantur per pendiculares in hanc diametrum: eæ omnes a circumferentia ellipfeos fecabuntur in eandem proportionem.

Ex l. 1. Apollonii Conicorum pag. XXI. demonftrat COMMANDINVS *in commentario fuper* V. *Sphæroideon* ARCHIMEDIS.

Sit enim circulus AEC. *in eo ellipfis* ABC *tangens circulum in* AC. & *ducatur diameter per* AC. *puncta contactuum,* & *per* H *centrum*. *Deinde ex punctis circumferentiæ* K. E. *defcendant perpendiculares* KL, EH, *fecta in* M.B. *a circumferentia ellipfeos. Erit ut* BH *ad* HE, *fic* ML *ad* LK. & *fic omnes aliæ perpendiculares*.

II.

⁕ Area ellipfis fic infcriptæ circulo, ad aream circuli, habet proportionem eandem, quam dictæ lineæ .

Vt enim BH *ad* HE, *fic area ellipfeos* ABC *ad aream circuli* AEC. *Eft quinta Sphæroideon* ARCHIMEDIS.

III.

Si a certo puncto diametri educantur lineæ in

68. Frontispiece to Kepler's *Rudolphine Tables* (1627). Kepler's tables, based on the observations of Tycho Brahe, brought a new, high level of accuracy to the predictions of planetary positions, thus attaining a goal of astronomers during the preceding millennia. Portrayed in the temple of astronomy, from left to right, are the astronomers Hipparchus, Copernicus, Tycho Brahe, and Ptolemy; a Chaldean stands in the background. Various astronomical instruments hang from the columns. At the very top, the imperial eagle dropping thalers from its beak symbolizes Emperor Rudolph II's financial support. On the dome are six goddesses, each representing an important aspect of Kepler's scientific work. The two figures at the left, one holding a globe casting a shadow and the other a telescope, represent Kepler's great work in the science of optics applied to telescopes and optical astronomy. The next goddess symbolizes logarithms and holds two rods, one of which is twice the length of the other, representing the natural logarithm of 1/2, which is written on her halo. Next is geometry with a square, a pair of dividers, and a diagram of the ellipse. The goddess with the unequal arm balance, or steelyard, symbolizes Kepler's law of areas; and the last one represents magnetism, the force that Kepler believed to emanate from the sun and to control planetary motions. The seated figure on the base of the temple is Kepler himself, and the banner next to him carries the titles of four of his books.

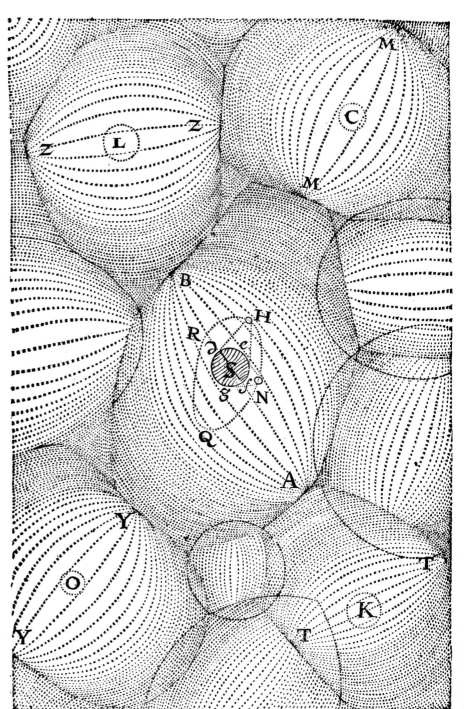

69. Celestial vortices. In the seventeenth century, a generally accepted explanation of the universe was based on the Cartesian system of vortices, huge whirlpools of tenuous or ethereal matter that were supposed to fill all of space. The vortices would carry the planets and their satellites around, thus producing the celestial motions. The sun *S* stands in the midst of the central vortex *AYBM*, surrounded by contiguous vortex systems with centers *C, K, O, L*. This system was expounded in detail in René Descartes's *Principia philosophiae* (1644), from which this illustration was taken.

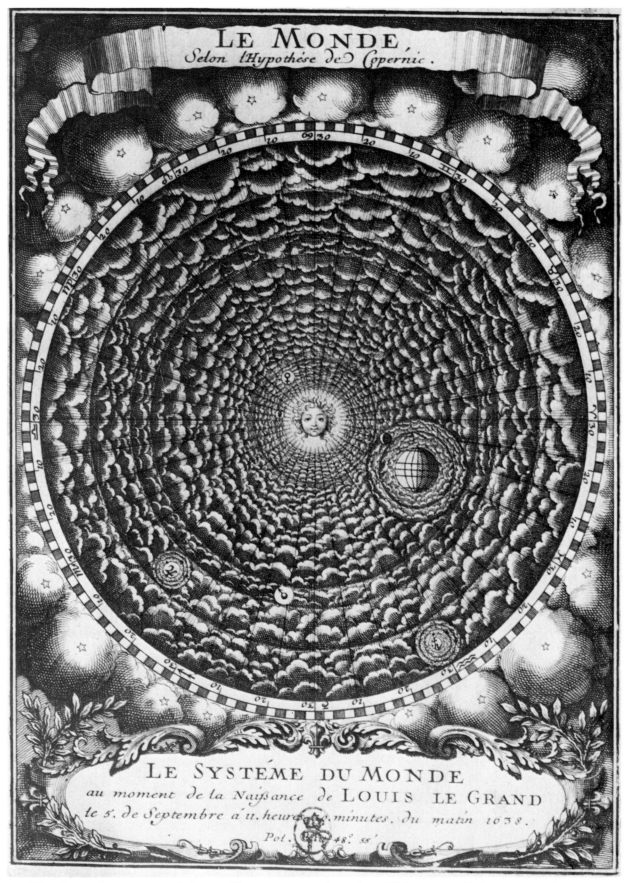

70. The Copernican system of the world as a system of vortices. This very realistic seventeenth-century engraving shows the sun and planets in the configuration of the Copernican system, but the planets and their satellites are obviously part of a system of swirling vortices. Additional stellar vortices surround the solar system. The print displays the aspects of the heavens at the time of the birth of Louis XIV, the "Sun King." It thus combines Copernican astronomy with Cartesian physics and astrology.

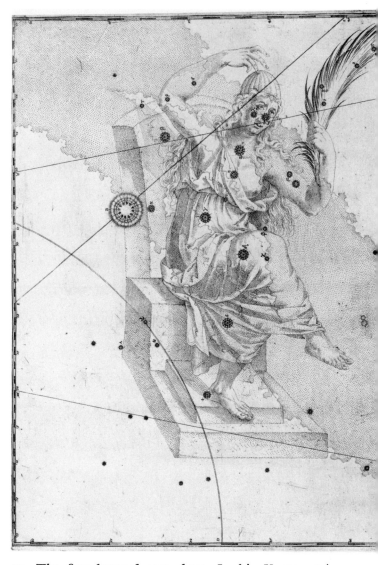

ea quam dixi annuli inclinatione , omnes mirabiles Saturni facies ſicut mox demonſtrabitur, eo referri poſſe inveni. Et hæc ea ipſa hypotheſis eſt quam anno 1656 die 25 Martij permixtis literis una cum obſervatione Saturniæ Lunæ edidimus.

Erant enim Literæ a a a a a a a c c c c c d e e e e g h i i i i i i l l l l m m n n n n n n n n n o o o o p p q r r s t t t t u u u u u; quæ ſuis locis repoſitæ hoc ſignificant, *Annulo cingitur, tenui, plano, nuſquam cohærente, ad eclipticam inclinato.* Latitudinem vero ſpatij inter annulum globumque Saturni interjecti, æquare ipſius annuli latitudinem vel excedere etiam, figura Saturni ab aliis obſervata , certiuſque deinde quæ mihi ipſi conſpecta fuit, edocuit: maximamque item annuli diametrum eam circiter rationem habere ad diametrum Saturni quæ eſt 9 ad 4. Ut vera proinde forma ſit ejuſmodi qualem appoſito ſchemate adumbravimus.

Cæterum obiter hic iis reſpondendum cenſeo, quibus *Occurritur iis quæ de annulo objici poſſent.* novum nimis ac fortaſſe abſonum videbitur, quod non tantum alicui cæleſtium corporum figuram ejuſmodi tribuam, cui ſimilis in nullo hactenus eorum deprehenſa eſt, cum contra pro certo creditum fuerit, ac veluti naturali ratione conſtitutum, ſolam iis ſphæricam convenire, ſed & quod annulum

71. The ring of Saturn. Galileo's telescope revealed that Saturn had a pair of protuberances. Astronomers wrestled with the problem of understanding the planet's shape without much success until the mystery was solved by Christiaan Huygens, who elucidated the nature of the phenomenon in 1659. When Huygens first discovered that Saturn is surrounded by a large ring, he could not bring himself to simply announce such an unbelievable discovery to the world. Accordingly, his 1656 announcement took the form of a published cipher. Later, after many more observations had shown without any possibility of doubt that Saturn is indeed surrounded by a ring, Huygens published an account of his discovery and called it *System of Saturn* (1659). The book contains diagrams of the ring as seen at various inclinations. On the page shown here, Huygens has reprinted the cipher and given the key to it: Saturn "is surrounded by a ring that is thin, flat, nowhere attached [to the planet], and inclined to the ecliptic."

56

72. The first lettered star chart. In his *Uranometria* (1603) Johann Bayer attempted to make a positive and unambiguous identification of every star visible to the naked eye. To each star in a constellation he assigned "one of the twenty-four letters of the Greek alphabet." For constellations with more than twenty-four stars, he used in addition the Latin alphabet. In his charts he not only gave the stars an alphabetical designation, but also included the traditional numeration of the stars in each constellation and the different names given them by Ptolemy and later astronomers. Thus, he hoped it would be easy to correlate the stars in his catalog with earlier records of them. This chart is especially noteworthy because it includes the supernova, or great new star of 1572, which was studied by Tycho Brahe. Tycho found the star to exhibit no observable parallax (shift in position relative to other stars when viewed from different places), which would result if it were near the earth. Hence, he concluded that it must be far out in the celestial spaces —in the realm of the fixed stars—where, according to the Aristotelian theory, no change can ever occur.

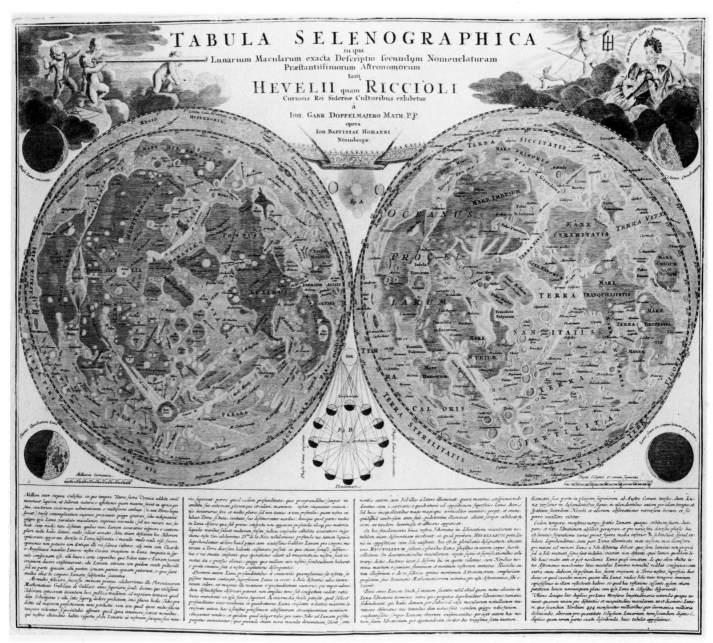

73. Maps of the moon. In his day, Johannes Hevelius was particularly famous for the extraordinary lunar maps he drew, engraved himself, and published under the title of *Selenographia* (1647). Among other achievements, Hevelius also provided many of the lasting names for lunar mountains, craters, and other formations. His drawings of the moon in different states of libration were another major contribution, as was the groundwork he laid in determining a system of lunar coordinates. Hevelius's moon maps were reproduced with comments in Giovanni Battista Riccioli's *Almagestum novum* (1651), and were later incorporated into Johann Gabriel Doppelmayr's *Atlas novus coelestis* (1742), from which this illustration is taken. At the bottom center, a diagram shows how the relative positions of the sun and moon produce the phases of the moon.

74. The solar system revealed by the telescope. A great classic scientific exposition that was popular in the early years of the eighteenth century was Fontenelle's *Conversations on the Plurality of Worlds* (first published in 1686). In one illustration from the book, a teacher learned in astronomy is discussing the heavens with a countess. The picture is particularly interesting because it portrays many revolutionary astronomical discoveries of the period. At the center of the sky, the sun appears as a spotted fiery object, radiating in all directions and illuminating the planets in their orbits around it. Each planet is an opaque sphere that casts a shadow on the side hidden from the sun. The earth is encircled by a single moon, and Jupiter is made distinctive by the four moons that Galileo observed. Saturn, in the outermost orbit, is surrounded by its ring, and near it are the satellite discovered by Huygens and the four additional satellites found by Gian Domenico Cassini.

75. An eclipse of the moon observed in Russia. In the sixteenth century, eclipses of the sun and moon were still spectacular events, arousing the interest of astronomers and the populace alike. One aim of almanac-makers was to provide accurate predictions of such phenomena. This illustration of a lunar eclipse is taken from a Russian manuscript of the sixteenth century.

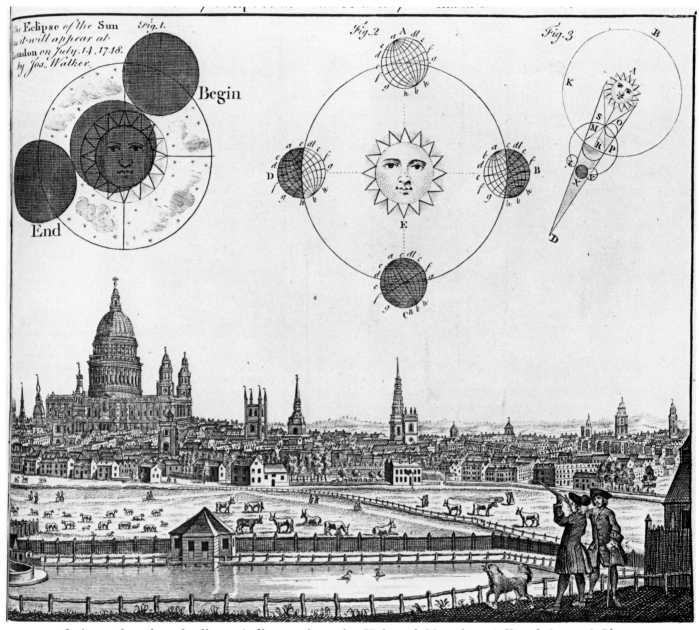

The Eclipse of the Sun as it will appear at London on July 14. 1748. by Jos. Walker.

Fig.1

Begin

End

Fig.2

Fig.3

76. An explanation of eclipses. A diagram from the *Universal Magazine of Knowledge and Pleasure* (1748) explains how eclipses occur in relation to the motion of the earth and moon. In this London scene, St. Paul's Cathedral is visible at the left.

77. Halley's predicted path of the moon's shadow during the eclipse of 1715. Halley wanted people to be apprised of the coming total solar eclipse "on the 22d Day of April 1715 in the Morning," since otherwise they might "be apt to look upon it as Ominous, and to Interpret it as portending evill to our Sovereign Lord, King George and his Government, which God preserve." Halley explained that the event is a "natural" occurrence, being "no more than the necessary result of the Motions of the Sun and Moon," and he especially asked observers to make note of "the duration of Total Darkness with all the care they can," so that the "dimensions of the Shadow" could be determined. As a result, he said, "we may be enabled to Predict the like Appearances for the future to a greater degree of certainty."

78. A prediction of a solar eclipse verified. After the solar eclipse of 22 April 1715 actually followed the predictions exactly, a second chart was published to emphasize this fact. It provides impressive evidence of the powers of astronomy to make exact and verifiable predictions. Halley could observe that "tho' our numbers pretend not to be altogether perfect," the predicted occurrence, path, and duration were so close to what actually occurred that "the correction they need is very small."

60

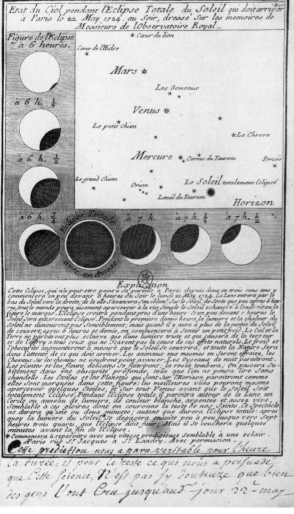

79. French verification of the 1724 eclipse. In contrast to the preceding pair of charts, this one pictures the exact appearance of the 1724 total eclipse of the sun at different times, rather than showing the path of the eclipse and the region of totality, as Halley had done. The reader was told that a similar eclipse had not been seen in Paris for 300 or 400 years. Describing the sequence of appearances, the author paid special attention to the solar corona, which, he said, would appear for one or two minutes at totality and would resemble the halo of saints. At the bottom of this chart an observer has written: "This prediction was verified by us as to the time, the duration, and other respects, which has persuaded us that this science [of astronomy] is not as doubtful as many people have believed to this day, the 22nd of May."

80. The comet of 1742. In 1705 Halley published his "Synopsis of the Astronomy of Comets," predicting that comets move in periodic orbits and return to view at stated times. He boldly declared that the comet of 1682 has a period of almost seventy-six years and that it would return to view at the end of 1758. When the comet appeared again at Christmas 1758, right on schedule, it was a great verification of Halley's theory. By the middle of the eighteenth century, all the appearances of comets aroused great astronomical interest. This illustration shows a comet as it passed over a building in the monastery of Einsiedeln in Zurich on the eve of 10 March 1742.

R. P. MAXIMILIANUS HELL e S. J.

81. Father Hell observing the 1769 transit of Venus. Father Maximilian Hell, a Viennese Jesuit, observed the transit of 1769 on the island of Vardö under the sponsorship of Christian VII, king of Denmark and Norway. Hell—accompanied by a fellow Jesuit, Father Johann Sajnovics—remained on Vardö for eight months collecting data on Arctic geography, meteorology, anthropology, and the language of the region. He also made a study of the aurora borealis and of barometric fluctuations. Later, Father Hell was associated with Franz Anton Mesmer in the doctrine of "animal magnetism." The transits of Venus across the face of the sun in 1761 and 1769 inspired the first large-scale international scientific undertakings because scientists recognized that such transits might provide a means for determining with great accuracy the scale of the universe. The method actually used was perfected by Halley. At least 120 observers at 77 different stations made observations of the transit in 1769, including Captain James Cook, who was in Tahiti.

SAPIENTISSIMI OPUS.

82. An eighteenth-century modification of the Cartesian vortices. Leonhard Euler, one of the greatest mathematicians and mathematical physicists of the eighteenth century, was very much concerned with problems of astronomy. One of his major contributions in this area lay in his study of the motion of the moon, particularly of the effects of solar perturbations on lunar motion, a problem that had baffled Newton. In addition to pursuing his mathematical studies, Euler also conceived a number of mechanisms to explain observed motions in the solar system in a Cartesian manner, that is, in terms of circular motions in the ether. He rejected the notion that empty space can exist, or that one body can act upon another at a distance without the intervention of something physical between them.

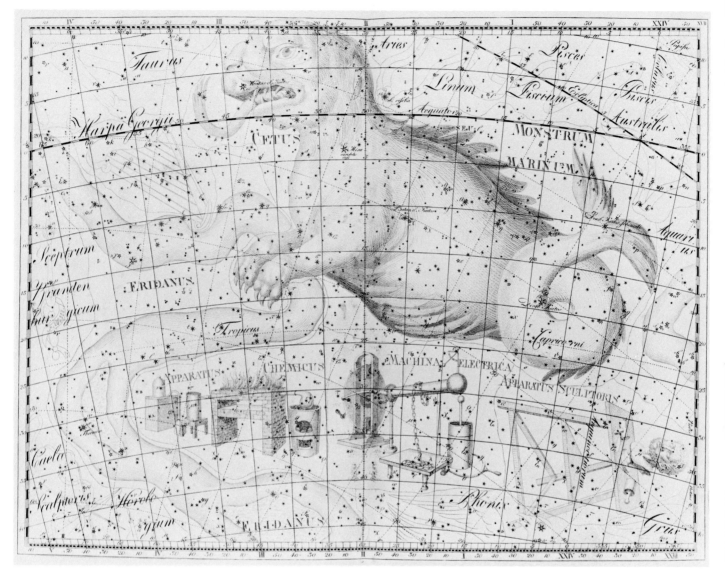

83. Fanciful constellations in the southern hemisphere. As geographical explorations made astonomers familiar with the stars of the southern hemisphere, an attempt was made to introduce names for newly discovered constellations based on the scientific and technological innovations of the modern age. J. E. Bode's celestial atlas *Uranographia, or Description of the Stars* (1801) introduced new constellations with such names as *Apparatus Chemicus* (chemical apparatus), *Machina Electrica* (electrostatic generator), and *Apparatus Sculptoris* (workbench of the sculptor). The electrostatic generator shown is a late eighteenth-century plate machine, connected to a long prime conductor with a quadrant electrometer at its top and a Leyden jar or capacitor beneath it.

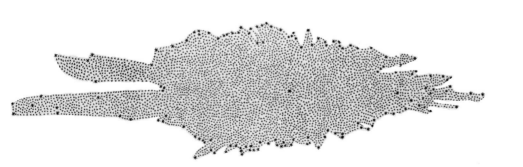

84. Our galaxy. In 1784 William Herschel published a famous paper, *On the Construction of the Heavens.* After studying a number of galaxies, or "island universes," and measuring the density of stars in various regions of the heavens, Herschel concluded that it is possible to determine the shape of our own galaxy. To a twentieth-century critic, what may be most interesting about Herschel's diagram is that he located our sun and earth at the center of the galaxy, "although perhaps not in the very centre of its thickness."

Fig. M

85. Hevelius and his wife observing the heavens together. Johannes Hevelius of Danzig (now Gdańsk), Poland, was one of the great astronomical observers of the seventeenth century. In 1673 he published an account of his observations, together with many illustrations of the instruments that he designed and used. One of the arresting features of this illustration is that Hevelius is shown making observations in collaboration with his wife. The role of women in astronomy has not fully been appreciated, although it can plainly be read in the record. Catherina Elisabetha Koopman Hevelius was not only a co-astronomer, but she was also hostess to many visiting astronomers, including Edmond Halley, and she edited many of her husband's unpublished writings after his death.

CHAPTER

7

The New Tools of Astronomy

Although Galileo and Thomas Harriot had applied the telescope to study the heavens in 1609, the instrument really came of age in the late seventeenth and eighteenth centuries. By the 1670s, when Newton made his discoveries concerning light and color, he realized that refracting telescopes of the times had definite limitations. They were constructed with two or more separate lenses, and they consequently suffered from chromatic aberration. This phenomenon arises from the fact that when the light from any star or other celestial object is passed through a simple lens, the several colors within the light come to a focus at different points. Not until well into the eighteenth century did scientists discover how to combine lenses in pairs to form an "achromatic" combination, in which each of the two combined lens elements was made of a different kind of glass in order to minimize chromatic aberration.

Newton's solution to the problem was to design a different kind of telescope in which the primary magnifying or light-collecting element is a concave mirror rather than a lens. Since a mirror reflects light, the problems of chromatic aberration are avoided. By the end of the eighteenth century, William Herschel had designed and constructed a huge reflecting telescope on the Newtonian model, with a mirror some 4 feet in diameter.

To get increasingly larger magnifications, other astronomers of the seventeenth century designed refracting telescopes that became longer and longer. Extremely long telescopes are a pronounced feature in pictures of observatories and astronomical instruments of the period. But by the end of the century, astronomers had found that smaller telescopes with mirrors would provide the same effect.

The accuracy of the new telescopes, eventually those with micrometers for exact measurements, developed in the seventeenth century, put stronger demands than ever on astronomical theories. Valuable products of these refinements were the star charts and tables assembled by men like John Flamsteed and Gian Domenico Cassini.

The seventeenth and eighteenth centuries were witness to the establishment of observatories. The first great observatory of modern times was the Royal Greenwich Observatory, established in 1675 by Charles II, who was particularly interested in the possible use of astronomy for navigation. In those days, the outstanding scientific problem of a practical kind was to discover a method for determining the longitude at sea. It was easy enough to find the latitude, since a navigator could readily determine the altitude or elevation of the pole—using a method that had been known since the time of Ptolemy, in the second century A.D. But finding the longitude was more difficult, and no satisfactory means for doing so had been discovered by the late 1600s. A navigator would be able to determine his longitude by finding the difference between his local apparent time and the time at some standard, say Greenwich, if he had a means of knowing the Greenwich time at the moment when he would determine the local apparent time. In the seventeenth century it was hoped that accurate tables of the motions and eclipses of the moons or satellites of Jupiter could be used in conjunction with observations of the eclipses of these satellites, in order to provide information on the difference in time or the longitude. This method proved to be impractical for shipboard navigation because it was impossible to see Jupiter's satellites on a tossing sea. Another proposed method was to compare observations of our moon with standard tables, but this method required accurate tables of the moon's motion, which were not available until the end of the eighteenth century. By then another solution had been found—a new kind of mechanical clock, the marine chronometer, which would give the navigator the Greenwich time and which he could then compare to his own local time.

The Royal Greenwich Observatory, the oldest existing government-financed scientific institution, was soon followed by the Paris Observatory, under the patronage of Louis XIV. Before long, there were other well-equipped public and private observatories all over Europe, including Russia, even though building them was a costly affair.

86. Hevelius's astronomical observatory. There seems to be no doubt that Hevelius had the finest astronomical observatory in Europe during the middle of the seventeenth century. It was built on a large platform that stretched across the roofs of three houses. Careful attention was paid to supporting systems so that the instruments would not be buffeted by the wind; and a rotating pavilion protected the larger ones from the weather. A 6-foot sextant and an azimuthal quadrant were included among the astronomical equipment. The three buildings below the platform, owned by Hevelius, housed a considerable library, workrooms, and a private printing press.

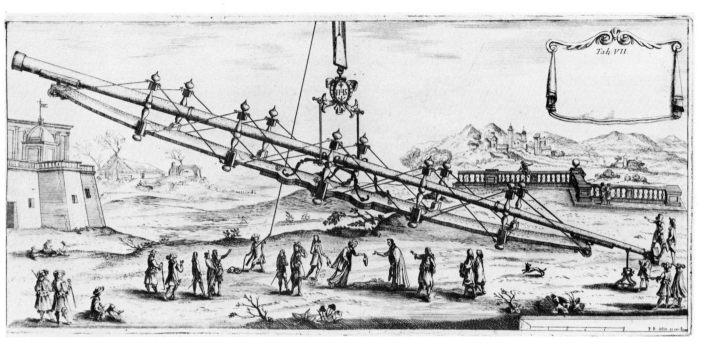

87. Campani's telescope. Giuseppe Campani, one of the most skillful designers and makers of telescopes in the late seventeenth century, made this enormously long example for an observatory in Rome. A rugged pulley system was needed to hoist the instrument in place. Note the external bracing, which reduced flexure of the telescope tube. He also made a 17-foot telescope used by Gian Domenico Cassini to study the sun's rotation and a 34-foot telescope that was presented by Louis XIV to the newly founded Royal Observatory in Paris. In addition, Campani invented the "composite-lens eyepiece" (one of his telescopes had a triple ocular plus an objective—four lenses in all), devised new methods of grinding lenses, made microscopes, and made observations of Saturn and of the spots on Jupiter.

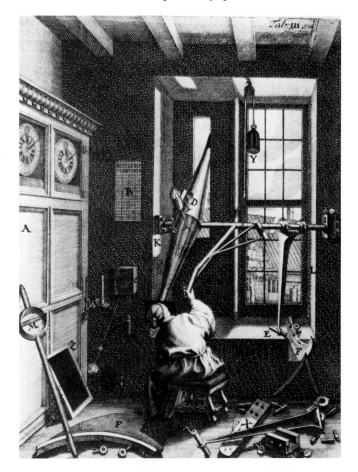

88. Rømer's transit instrument (1684). Ole Rømer, a Danish astronomer, devised many new astronomical instruments, some of which were adaptations of the telescope to pre-telescopic mountings. Here he is seen observing transits with his telescope set in the plane of the meridian. The time of transit is determined by the clocks visible on the left wall and a pendulum device on the wall at the left of the observer. One of the interesting features of Rømer's telescope is the handle that he is using to control the angle of elevation. Note that a pulley-and-weight system compensates for any flexure of the telescope's support caused by the weight of the instrument. Two of Rømer's most well-known inventions are the altazimuth and the equatorially mounted telescope. He was also the first to determine that light had a finite speed. Rømer did this by noticing that apparent irregularities in the periods of Jupiter's satellites could be correlated with Jupiter's position, and he correctly concluded that it requires time for light to travel from Jupiter to us.

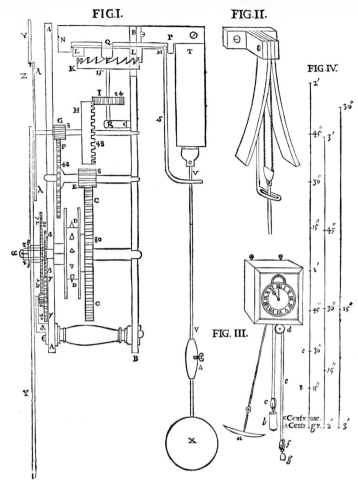

FIG.I.   FIG.II.

FIG.IV.

FIG.III.

89. The pendulum clock. In the seventeenth century an accurate clock was needed to time astronomical observations and serve in determining longitude at sea. Galileo had discovered the isochronism of the pendulum in the late 1500s and worked out the design of a clock regulated by a pendulum, but the clock was apparently never constructed. This clock of Huygens had two main features that he discovered to be necessary for uniform timekeeping: a pendulum connected to an escapement mechanism that regulates the motion of the wheelwork (Fig. I, K), and two metal strips shaped in the curve of an epicycloid to check the swing of the pendulum (Fig. II). There were rival claims to parts of the invention, notably by Robert Hooke, but the fact remains that in 1673 Huygens published the first complete treatise on the modern clock (from which this illustration is taken), and early clocks were actually made according to his design. Fig. III shows the clock with the weight that drives it and the wide swing of the pendulum that requires the checks.

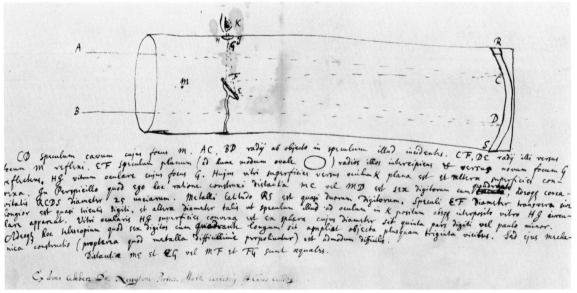

90. Newton's drawing of his reflecting telescope. Isaac Newton's first publication (1672) was devoted to experiments on light and color that convinced him it would be impossible to produce refracting telescopes (composed of lenses) without chronic aberration. Accordingly, Newton invented a telescope in which magnification is produced by a concave reflecting mirror, rather than by a lens. In this previously unpublished drawing of such a telescope, the mirror is located at the right end of the barrel, between letters *R* and *S*. A second small mirror, set to the left of it at a 45° angle in the center of the tube, reflects the gathered light at right angles into an eyepiece through which the astronomer makes his observations. A note at the bottom states that this drawing was "the gift of the famous Mr. Newton, Prof. of Math. at Cambridge, of Caius Colledg." Newton, however, had been a fellow of Trinity and not Caius. The note is in the handwriting of Johann Jacob Huber, a Swiss astronomer who worked in England in 1754–1755 with James Bradley, the astronomer royal. Huber's note does not say to whom Newton made this gift, but perhaps it was to Bradley, who may in turn have given this drawing to Huber.

PROSPECTUS INTRA CAMERAM STELLATAM.

91. The octagon room. In the octagonal clock room of the Royal Greenwich Observatory, founded in 1675, John Flamsteed, the first astronomer royal, made observations of the moon and the stars that were later incorporated in his famous star catalog, the *Historia coelestis Britannica* (1725). Two royal portraits, those of Charles II and James, Duke of York (later James II), hang above the door. Like the rest of the observatory, this room was designed by the scientist-architect Christopher Wren. An interesting feature, at right, is the ladder-like device Flamsteed used to support his long telescopes. The room incorporated two clocks, designed by Thomas Tompion, which could run for a year without interruption and had 13-foot-long pendulums. They were used by Flamsteed to determine whether the rotation of the earth on its axis is constant, that is, whether the rotating earth is a good clock, a fact of considerable significance for ascertaining the longitude. This room was also used for observations of solar and lunar eclipses, comets, occultations of stars by the moon, eclipses of Jupiter's satellites, and other phenomena. Since the telescopes were not permanently fixed to the floor, they could be moved from one window to another as needed.

92. The Royal Observatory at Paris. Construction of the Paris observatory was begun in 1667. As evident in this seventeeth-century print, it was a center of great activity, with telescopes set up on the grounds of the observatory as well as on the roof. The Invalides can be seen in the distance at the right.

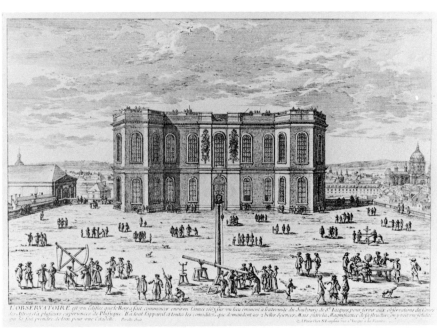

L'OBSERVATOIRE est un Edifice que le Roy a fait construire environ l'anné 1667...

93. The first director of the Paris observatory. Gian Domenico Cassini was an astronomer in his native Italy prior to becoming the inaugural director of the Royal Observatory in Paris in 1669. In this engraving by L. Cossin (or Coquin), Cassini is seen holding a small telescope and pointing to the new observatory building. He introduced telescopes of long focal length that required elaborate supports, and pioneered in adopting the newly invented micrometer. One of the telescopes was equipped with a lens having a focal length of 136 feet. Cassini made many astronomical discoveries, among them the second, third, fourth, and fifth satellites of Saturn. In particular, he developed tables of the satellites of Jupiter that were useful for navigators and geographers. He also participated in the determination of the solar parallax (yielding the distance of the sun from the earth) and began a careful measure of the shape of the earth, hoping to support his erroneous view—opposed by Huygens and Newton—that the earth is prolate (flattened at the equator) rather than oblate (flattened at the poles).

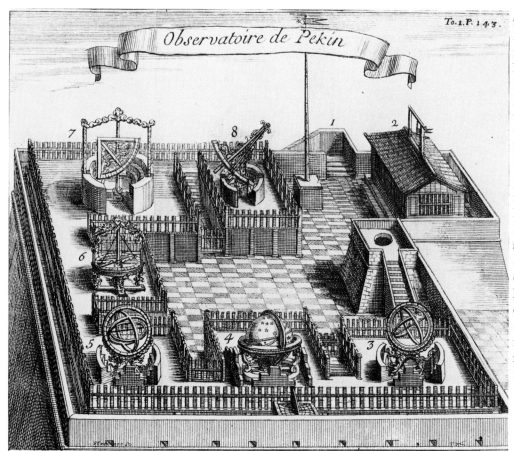

Observatoire de Pekin

To. 1. P. 143.

94. The Peking Observatory. In 1674 Melchior Haffner made an engraving of the Peking Observatory for the book by Ferdinand Verbiest, a Belgian Jesuit, who refitted the installation. The illustration was reprinted many times, and this copy is taken from a French work on China, published in 1696. The instruments pictured are very similar to those being used in Europe at the time. These are an equatorial armillary sphere (3), a celestial globe (4), an ecliptic armillary sphere (5), a horizon circle (6), a huge quadrant (7), and a sextant (8). Both the observatory and its instruments have survived more or less intact into the twentieth century.

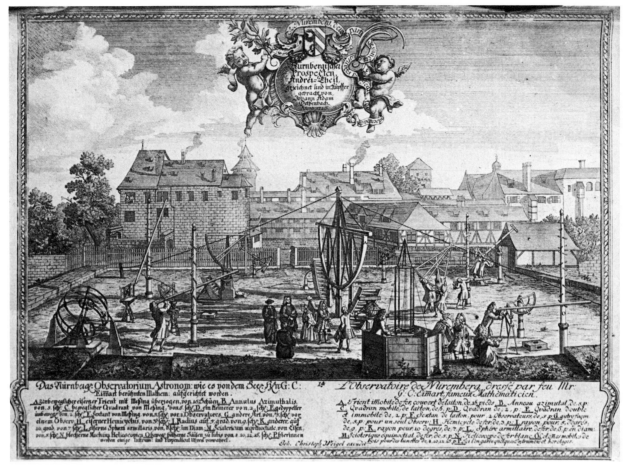

The Nuremberg Observatory in 1716.

95. The Nuremberg Observatory in 1716. Georg Christoph Eimmart, an artist and copperplate engraver, was also an amateur astronomer, who, in the late seventeenth century, established the private observatory seen here. At the center is a huge double quadrant that was 16 feet across, and at the far left is a 6-foot-high armillary sphere. This observatory became one of the highlights of Nuremberg frequented by fashionable ladies and gentlemen.

96. An eighteenth-century observatory in St. Petersburg (Leningrad). Seen from the river, the observatory in the tower of the *Kunstkammer* (a natural history museum including a cabinet of natural curiosities plus a collection of scientific instruments) of the St. Petersburg Academy of Sciences is plainly visible. In the distance is the famous fortress of Peter and Paul.

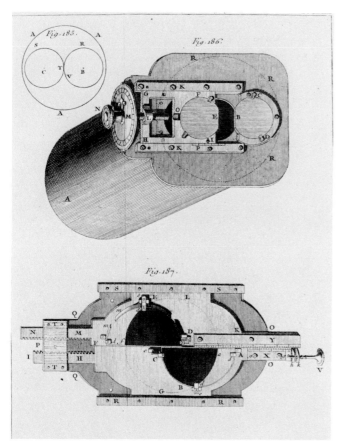

98. Herschel's telescope of 1789. William Herschel's telescope was the largest ever built up to that time. It was 40 feet long and had an aperture of 4 feet. Herschel, a German musician who came to England and became the most famous observational astronomer of his age, was assisted in all of his observations by his sister Caroline. His discoveries included many new phenomena, among them the revolution of the planets, properties of nebulae, aspects of double stars, and new satellites of Saturn. Herschel is most famous for the discovery of a new planet, which he named "Georgium sidus" (or "George's star"), in honor of George III, and which is now known as Uranus.

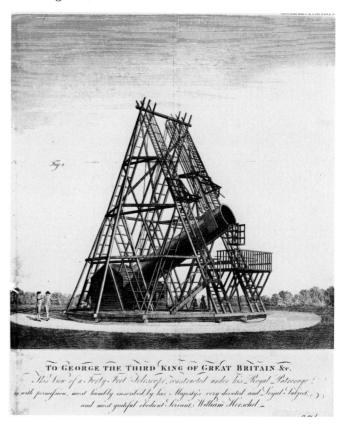

97. The micrometer. The precision of telescopic measurements depended on the introduction of the micrometer in the late seventeenth century. This device, which was put into the eyepiece of a telescope, made it possible to measure either very small angles (or distances between objects) or the diameters of very small objects, such as planets and their satellites. The first micrometer described in print (by Huygens) consisted of a tapered graduated wedge that could be moved until it just covered the object to be measured. Other micrometers used either parallel threads, one fixed and the other movable, or closely spaced grids of wires or threads. This illustration, taken from Joseph Lalande's popular *Traité d'astronomie* (1764), shows two forms of micrometer commonly used in the eighteenth century. The upper one, invented by Pierre Bouguer, is an object-glass micrometer. The lower one, invented by John Dollond, has a divided object glass; when the two parts of the glass were placed together the observer saw a single image, but when they were separated (as shown in the diagram) each half produced its own image and the distance between them could be determined from a scale and vernier. The original concept of dividing the image in this way goes back to Rømer.

Part Three

THE
EXACT
SCIENCES

99. A contest between the old and the new arithmetic. The mathematician on the left (symbolized by Boethius) is using the new Hindu-Arabic numerals, whereas the one on the right (symbolized by Pythagoras) is still reckoning with counters on a reckoning board. Our modern word "bank" with its "counters" derives from the old Germanic word *banc* or *bank* (via the French *banque* and Italian *banca*), for a money changer's bench or table; the word "bankrupt" comes from *banca rotta*, a broken bank. From the expressions on the faces of the two contestants, it is easy to see which one is winning. The new numerals greatly simplified scientific computations.

8

The Beginnings of Modern Mathematics

During the sixteenth and seventeenth centuries, mathematics was studied for its practical uses as well as for its value in teaching the principles of logic and reasoning. The chief applications of mathematics were to astronomy, geography, navigation, surveying, and commerce.

Many works explained how to do practical calculations, using the counters (counting pieces, shaped like thick coins, which were moved along lines on a counting bench or "bank," much as in an abacus), as well as the more familiar methods of addition, subtraction, multiplication, division, and the extraction of roots. But the full advantage of the new Hindu-Arabic numerals was not obtained until decimal fractions replaced traditional sexagesimal fractions (based on the number 60) in the late sixteenth century, largely owing to the efforts of Simon Stevin. From the sixteenth century on, elaborate trigonometric tables were produced, primarily for astronomical computations. Such computations were very laborious, until John Napier

simplified them by his invention of logarithms in 1590. In 1621 William Oughtred invented the slide rule to perform simplified mechanical computations, using a pair of logarithmic scales.

From a twentieth-century point of view, it is fascinating to discover digital machines invented to perform calculations as early as the seventeenth century. In 1624 Wilhelm Schickard designed and constructed the first machine to perform digital calculations mechanically. Unfortunately, the only example that he built was destroyed in a fire, so the world did not learn about his invention until the mid-twentieth century, when some of Schickard's letters, including a description and a drawing of the machine, came to light. Apparently, he had invented the machine to save his friend Kepler from the labors of tedious calculations. The direct line of descent of modern mechanical digital calculating goes back only as far as Blaise Pascal. In order to assist his father, who was a tax collector, in 1645 Pascal invented an ingenious machine for adding and subtracting. Altogether, some twelve machines were built for Pascal, a number of which still survive and are obviously efficient.

Later in the seventeenth century, technology took a great leap forward when Leibniz produced a calculator that could perform multiplication mechanically, by means of a kind of stepped drum or pinwheel. Leibniz well expressed the philosophy of those who have designed calculators or computers when he said, "It is unworthy of excellent men to lose hours like slaves in the labor of calculation which could be safely relegated to anyone else if a machine were used."

The greatest advance in mathematics in the centuries immediately following the beginning of the Scientific Revolution was the discovery or invention of the calculus—an achievement that should be attributed to the independent creative labors of G. W. von Leibniz and Newton. Leibniz, who published his work first, has the honor of being the one to make the calculus known to the world at large; but Newton's notebooks and private papers, which were circulated in manuscript among his circle of acquaintances, give him precedence. When Leibniz asked Newton for information concerning his discovery, Newton sent a reply chiefly in the form of a cipher, or code. In those days, it was common to announce in cryptic form an odd or unusual discovery that might be premature (see Ill. 110). But the use of a cipher was not the customary manner of sending information about new discoveries to a correspondent. There is no available information about Leibniz's attempts to decipher Newton's message.

Leibniz developed the common algorithm for the calculus with which we are familiar today, using d's in the notation, whereas New-

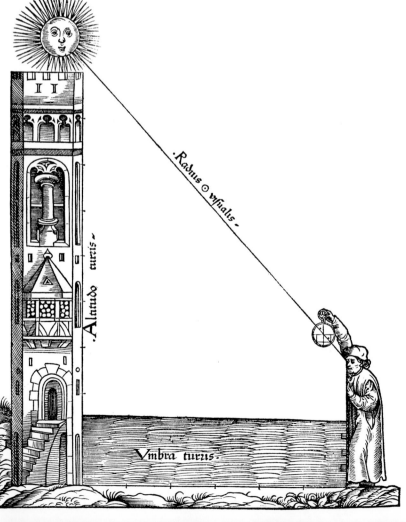

100. Using practical geometry to find the height of a tower. This illustration, taken from *An Explanation of the Construction and Use of the Astrolabe* (1513) by Johannes Stöffler, delineates the right triangle formed by a tower (actually the height of the tower diminished by the distance from the ground to the observer's eye), its shadow on the ground, and a line from the end of the shadow to the tower's top, toward the sun. Since the length of the shadow can be measured, the surveyor can readily compute the height of the tower by using an astrolabe. The observer stands at the edge of the shadow and aligns the pointer of the astrolabe with the sun, thus producing a right triangle on the face of the astrolabe that is similar to the physical triangle formed by the tower and its shadow. The height of the tower then becomes the only unknown term in a proportion of four terms. A correction then has to be made for the height of the observer's eye above the ground, which must be added to the result. This method works for a sighting made at any known distance from a tower. If the horizontal distance cannot be measured, however, the height can still be determined from two triangulations, made at two points whose distance from each other is known. The same principle could be used to find the distance from an elevated point on shore to a ship, given the height of the point.

ton's final notation was to put dots or "pricks" above letters. Newton's notation is also used today; both the Leibnizian and Newtonian notations appear in the famous equations of Joseph Louis Lagrange. Since the calculus is a mathematical way of describing functional relations of change, it has become the language of the exact sciences.

WEiter spricht man das ein stuck Buchsen auff zwen punckten gericht sey/ wann solches perpendicular oder faden der pleyschnur/ dises Jnstruments/ gantz gerad fallet auff die, lini des vnterschieds/oder abteilung des anderen puncten. Also wird auch solches stuck auff den dritten puncten gericht/wann die ge melt pleyschnur den dritten puncten/auff das eigentlichst betrifft/Also soltu ymer furt an verstehen ein yedes stuck auff den vierten / fünfften / sech sten /vnd ander volgende puncten zu richten/wie dir dise volgende figur anzeigt.

Wann

101. Tartaglia's gun quadrant. At right an artilleryman is using a type of quadrant with a plumb bob, devised by Niccolò Tartaglia to measure the angle of elevation. A long arm of the quadrant rests in the bore of the gun. A second artilleryman is measuring the range. This determines the angle of elevation to be used. Tartaglia explained that for maximum range the elevation should be 45°. He supposed that the trajectory of the bombard (medieval cannon) consists of three parts: a straight line extending from the center of the gun barrel, a short circular arc, and then a straight line downward. Ignoring the very small circular arc, the trajectory was considered as a right triangle. If the "force" of projection were known, it could then be considered equivalent to the length of the hypotenuse of the triangle, and the range could be computed from the angle of elevation.

77

mi.	Sinus	Logarithmi	Differentia	Logarithmi	Sinus	
0	1908090	16564818	16379581	185437	9816272	60
1	1910945	16549865	16363862	186003	9815716	59
2	1913800	16534935	16348365	186570	9815160	58
3	1916655	16520028	16332890	187138	9814603	57
4	1919510	16505144	16317438	187706	9814045	56
5	1922365	16490283	16302008	177275	9813486	55
6	1925220	16475445	16286600	188845	9812926	54
7	1928074	16460630	16271214	189416	9812366	53
8	1930928	16445837	16255849	189988	9811850	52
9	1933782	16431067	16240506	190561	9811243	51
10	1936636	16416320	16225185	191135	9810680	50
11	1939490	16401595	16209886	191710	9810116	49
12	1942344	16386895	16194610	192285	9809551	48
13	1945197	16372216	16179355	192861	9808986	47
14	1948050	16357559	16164121	193438	9808420	46
15	1950903	16342924	16148908	194016	9807853	45
16	1953756	16328311	16133716	194595	9807285	44
17	1956609	16313720	16118545	195175	9806816	43
18	1959462	16299151	16103395	195756	9806147	42
19	1962314	16284604	16088266	196338	9805577	41
20	1965166	16270079	16073159	196920	9805006	40
21	1968018	16255576	16058073	197503	9804434	39
22	1970870	16241095	16043008	198087	9803861	38
23	1973722	16226636	16027964	198672	9803287	37
24	1976574	16212198	16012940	199258	9802712	36
25	1979425	16197782	15997937	199845	9802137	35
26	1982276	16183388	15982955	200433	9801560	34
27	1985127	16169016	15967994	201022	9800984	33
28	1987978	16154665	15953053	201612	9800406	32
29	1990829	16140336	15938133	202203	9799827	31
30	1993679	16126028	15923233	202795	9799247	30

78

102. Logarithms. The first table of logarithms was published in 1614, with a brief description of their use by their inventor, John Napier. Napier's tables give logarithms to seven or eight places for every minute of angle from 0° to 90°, and make tedious astronomical calculations much simpler. This page contains the logarithms for the sine of 11°. In 1620 Kepler publicly thanked Napier for his invention and for simplifying the task of making astronomical calculations.

103. The slide rule. William Oughtred is generally considered to be the inventor of the slide rule, which is, essentially, an instrument with numbers set out on a circular or straight line scale, according to logarithmic values. By moving one scale along the other, lengths that correspond to logarithms can be added or subtracted from each other mechanically. These two operations correspond to multiplication by adding logarithms and division by subtracting logarithms. Oughtred's circular slide rule was described in 1632 in his *Circles of Proportion and the Horizontal Instrument.*

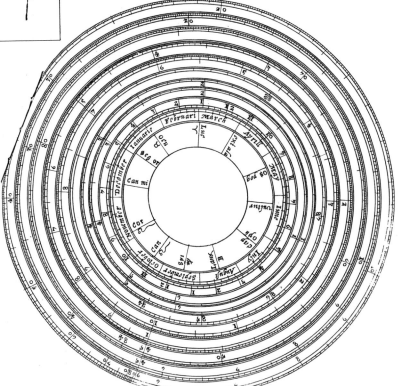

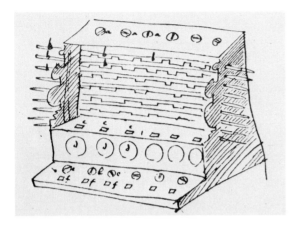

104. The first mechanical digital computer. To make Kepler's long and arduous computations easier, his friend Wilhelm Schickard of the University of Tübingen invented a mechanical calculator or calculating machine. This drawing of the device occurs in a letter sent by Schickard to Kepler in 1624. Numbers were entered in Schickard's machine by turning the dials in the base; the actual computations were then made by sliding the rods in the upper part, back and forth, thus mechanizing a set of Naperian computing rods. One example of the machine was constructed, but it was destroyed by fire before Kepler could use it. Schickard's invention remained wholly unknown until the account of it was discovered soon after the end of World War II.

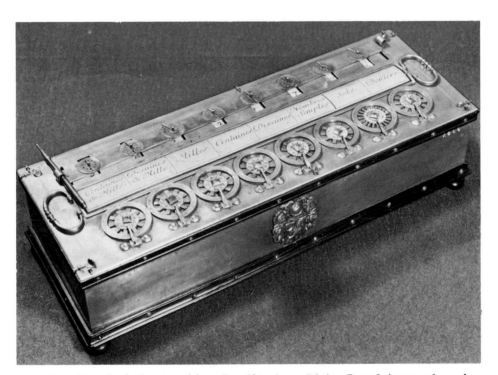

105. Pascal's calculating machine. In the 1640s, Blaise Pascal invented an ingenious machine that performed the functions of addition and subtraction. One of its novel features was the automatic "carry." When adding, say, 9 and 7, by turning a wheel through nine and then seven clicks, the "1" in the result, 16, would automatically be transferred to the appropriate register so that the correct answer could be read in the numbers that showed through the windows in the upper part of the machine. One of the most remarkable aspects of Pascal's device was the use of complements, so that subtraction could be performed by direct addition. A complement is the difference between 10 and an integer from 0 to 9. Thus 9 minus 6 is found by adding 4 (the complement of 6) to 9, to yield 13; the answer is 3. The wheels were turned by inserting the tip of the metal stylus found at the left side of the machine into the openings between the spokes of the large wheels.

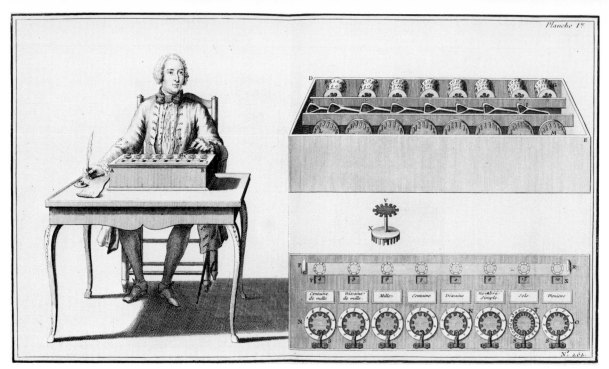

106. Pascal's machine in use. A composite illustration of the eighteenth century shows the construction of the toothed wheels (*upper right*) that perform the operations of addition and carrying. Each main wheel or register has two sets of numbers: the first, from 0 to 9 for performing addition; the second, from 9 to 0 for performing subtraction by adding complements. The lower diagram shows the wheels on which the operator enters the numbers in the individual registers. The two registers at the extreme right count to 20 and 12 rather than to 10, in accordance with the seventeenth-century French currency system in which *livres, sols,* and *deniers* were the equivalent of British pounds, shillings, and pence, with twenty *sols* in each *livre* and twelve *deniers* in each *sol.* The first six windows from left to right refer to *livres* and translate as: hundreds of thousands, tens of thousands, thousands, hundreds, tens, and integers. This breakdown indicates that the machine was designed for use in monetary reckoning (as in taxes) and not for pure arithmetic or science. If the sliding bar marked *PR* is moved forward, the openings for addition are covered up and those for subtraction are uncovered. The man at the left is reckoning accounts with the aid of the calculator.

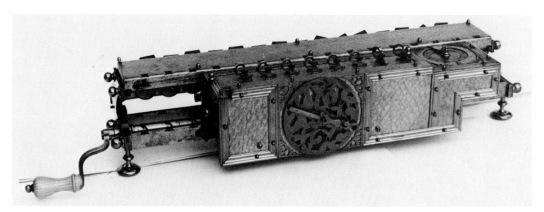

107. Leibniz's calculating machine. Leibniz invented two versions of a calculator, which were significant improvements on Pascal's creation. By means of pins of different lengths or a stepped drum, the new device could perform multiplication in a direct and easy manner. Leibniz also pioneered the development of binary arithmetic (an arithmetic using 2 as the number base rather than our common base of 10), which is standard for modern computers.

108. **The uses of mathematics.** The title page of *A New System of Mathematicks* (1681) illustrates the practical interests of the book's author, Sir Jonas Moore, Surveyor-General of the Ordnance. The lower part of the illustration displays various instruments used by practical mathematicians in surveying, measuring time, navigation, and astronomy. The assembled mathematicians are shown busy putting mathematics to use. It is obvious that the fields being stressed are astronomy, navigation, and geography. Moore was one of the primary forces behind the establishment of the Royal Observatory at Greenwich.

109. **Leibniz's method for solving maximum and minimum problems by means of the differential calculus.** A page from an article published by Leibniz in the *Acta eruditorum* (1684) displays his use of the familiar *d* notation in the algorithm that he devised for the calculus. Leibniz and Newton were rivals in the invention or discovery of the calculus, and it is generally agreed that both men have claims to independent creation, but Leibniz was the first to publish his method.

110. Newton's cipher for the calculus. A page from Newton's private *Waste Book* contains a record in his own handwriting, dated October 1676, of the code he used in reply to Leibniz's request for information about his work on the calculus. Here Newton explains how the letters 6 *a*'s, 2 *c*'s, *d*, *ae*, 13 *e*'s, 2 *f*'s, 7 *i*'s, and so on, could be arranged into words forming a sentence, which can be translated "For any equation with fluent quantities, to find the fluxion and vice-versa." In a second cipher, which follows the first, Newton went on to state the general principles of the calculus. "One method consists in extracting a fluent quantity from an equation at the same time involving its fluxion; the only other in assuming a series for any unknown quantity whatever, from which the rest could be conveniently derived." In the main part of his lengthy letter, Newton did show some of his methods to Leibniz.

111. Newton's first published paper on the calculus. Newton's *Treatise on the Quadrature of Curves* (1704) was printed as a supplement to his *Opticks*. On this page he indicated the first, second, third, and fourth derivatives (or "fluxions") by the use of one, two, three, or four dots above the letters z, y, x, and v, which stand for variables.

TRACTATUS

DE

Quadratura Curvarum.

Quantitates indeterminatas ut motu perpetuo crescentes vel decrescentes, id est ut fluentes vel defluentes in sequentibus considero, designoq; literis z, y, x, v, & earum fluxiones seu celeritates crescendi noto iisdem literis punctatis ż, ẏ, ẋ, v̇. Sunt & harum fluxionum fluxiones seu mutationes magis aut minus celeres quas ipsarum z, y, x, v fluxiones secundas nominare licet & sic dignare z̈, ÿ, ẍ, v̈, & harum fluxiones primas seu ipsarum z, y, x, v fluxiones tertias sic z⃛, y⃛, x⃛, v⃛, & quartas sic z⃜, y⃜, x⃜, v⃜. Et quemadmodum z, y, x, v sunt fluxiones quantitatum z̈, ÿ, ẍ, v̈, & hæ sunt fluxiones quantitatum z⃛, y⃛, x⃛, v⃛ & hæ sunt fluxiones quantitatum primarum z, y, x, v : sic hæ quantitates considerari possunt ut fluxiones aliarum quas sic designabo,

ż,

9

The Science of Statics

In the seventeenth and eighteenth centuries, the principles of statics were applied to the analysis of the five simple machines: the lever, wedge, wheel and axle, inclined plane, and the pulley. Galileo pioneered in the study of stresses and strains, or the science of the strength of materials, one of the "two new sciences," to which his last published great treatise was devoted. In 1586 a highly original analysis of the forces exerted on inclined planes was displayed by the Dutch engineer Simon Stevin of Bruges, who took advantage of the general principle of the impossibility of perpetual motion.

A powerful tool for the science of statics was the principle of the parallelogram of forces, providing a geometric way of combining the effects of two separate forces by finding a single force that would produce the same effect. Conversely, the parallelogram rule made possible the resolution of any given force into two components.

One of the important laws of statics, that of direct proportionality of stress and strain, was discovered by Robert Hooke in his experiments with springs. Of course, this law is valid only so long as the experimenter does not put too great a strain on the spring being used. If the spring's elastic limit is exceeded or if the spring is permanently distorted, it will be converted into a wholly different spring.

By the end of the eighteenth century, the subject of statics took a radically new turn, chiefly as a result of the experiments of Charles Augustin Coulomb, the same scientist who discovered the laws of electrical and magnetic force. Coulomb inaugurated the exact study of friction with new types of experimental devices, and he thereby converted a subject marked by murky discussions into a new branch of exact experimental science.

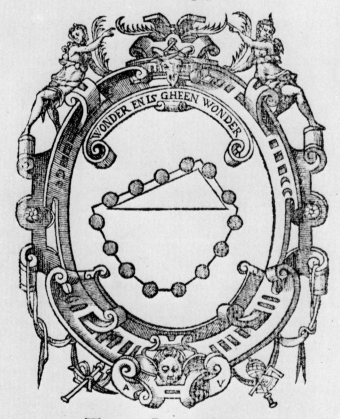

DE
BEGHINSELEN
DER WEEGHCONST
BESCHREVEN DVER
SIMON STEVIN
van Brugghe.

TOT LEYDEN,
Inde Druckerye van Christoffel Plantijn,
By Françoys van Raphelinghen.
cIɔ. Iɔ. LXXXVI.

112. The inclined plane. In 1586 Simon Stevin produced an analysis of the inclined plane using the idea of a chain, or wreath of equal spheres on a string, placed over two inclined planes, one twice the length of the other. Such a chain does not move, but stays in equilibrium, Stevin argued, since there can be no perpetual motion; accordingly, the downward pull of the four spheres on the left plane must be equal to, or must balance, the pull of the two spheres on the right plane. This happens because the right plane is steeper than the one on the left. Stevin's explanation is based on the principle of components of forces, which leads to the idea of the parallelogram of forces. The angle of inclination determines the effective component of the weight of the two balls on the right and that of the four balls on the left. In each case, the "effective component" of gravity is inversely proportional to the length of the inclined plane. The left component is measured by the weight of four spheres divided by a length of two units, while that of the right is measured by the weight of two spheres divided by a length of one unit. The two components are thus equal, producing equilibrium. The motto above the plane says, "The wonder is that there is no wonder."

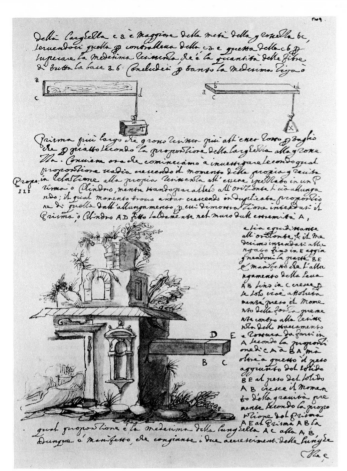

113. An analysis of strains and stresses in beams. A page from the manuscript of Galileo's book *Two New Sciences* (1638) depicts two aspects of his studies of the forces in beams under stress. The diagrams at the top of the page illustrate the greater strength (or resistance to flexure and breaking) of a beam placed on edge (*left*) compared to a flat beam (*right*). The figure at the bottom of the page demonstrates how a beam supported at only one end can be under strain and be liable to fracture; in this case, the beam's own weight in the section *DEBC* further from the support could cause the beam to fracture at *BD*.

114. The application of statics to the human body. Galileo's disciple, G. A. Borelli, used the principles of the analysis of forces to study the actual stresses and strains in the bones and muscles of humans and animals. On this page from his work *De motu animalium* (published posthumously in 1680 and 1681), he illustrated two studies of humans bearing different loads (5 and 6). Another pair of diagrams (9 and 10) shows how the actions of the muscles in the outstretched arm of a man holding a weight in his hand are similar to those in a pulley system. Figures 7 and 8 are also studies on various pulley configurations.

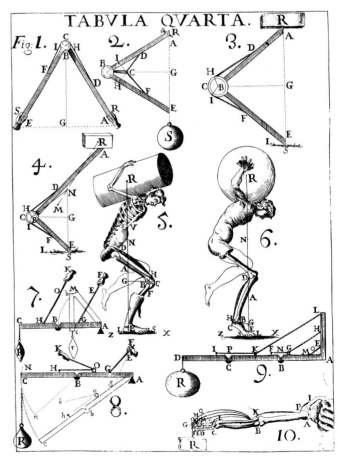

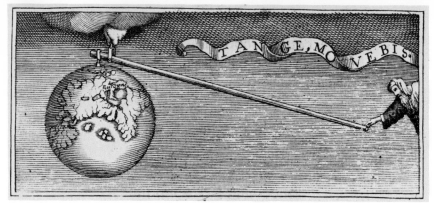

115. "Give me a place to stand and I will move the earth." Pierre Varignon illustrates (1687) his interpretation of the famous statement attributed to Archimedes. The banner reads *"Tange, movebis,"* or "Touch it and you will move it." Since Archimedes had discovered the law of the lever, there is a nice logic in Varignon's having used a lever to explain and illustrate how Archimedes might have thought the earth could be moved by a machine. It is more likely that Archimedes had in mind the enormous mechanical advantage possible from using a system of pulleys.

116. A seventeenth-century solution for moving the earth. Reflecting the seventeenth-century advances in physical mechanics, J. S. Delmedigo suggested that Archimedes had a system of gears in mind when he claimed it would be possible to move the earth. Delmedigo documented his thought in a Hebrew book on natural philosophy that was published in 1629 in Amsterdam.

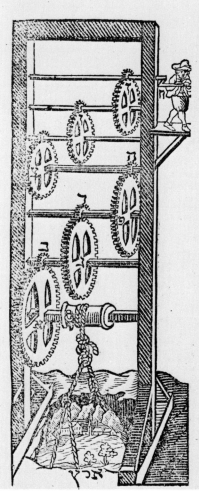

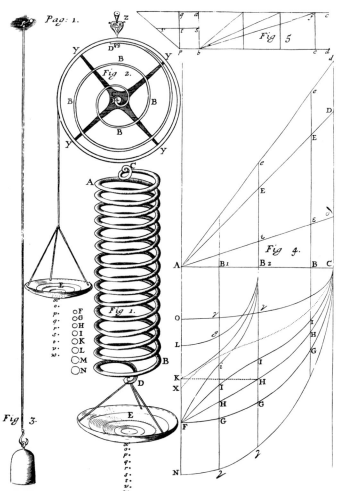

117. The law of the spring. Robert Hooke's *De potentia restitutiva; or Of Spring, Explaining the Power of Springing Bodies* (London, 1678) explained in detail the relationship between stress and strain in springs. Hooke produced a measurable stress by applying a weight to a straight wire (figure 3), to a heliacal spring (figure 1), and to a spiral spring (figure 2). In the graphs in the upper part of figure 4, he revealed the linear relationship between stress and strain. Hooke first announced his discovery in the form of an anagram: *ceiiinosssttuu* (one of the *v*'s may be written as a *u*), which stood for the words *"Ut tensio sic vis"* ("As the tension, so is the force"). In other words, there is a direct proportionality between stresses and strains.

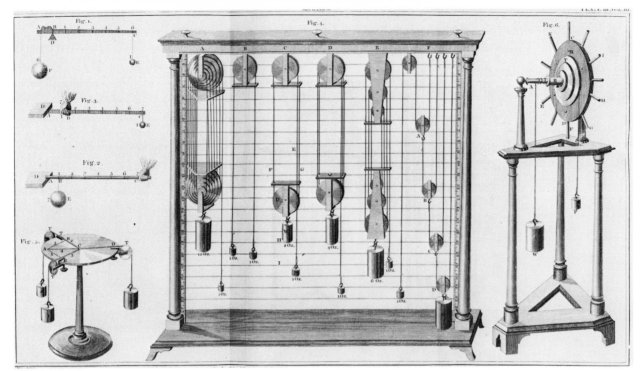

118. Three "simple machines." George Adams's popular work *Lectures on Natural and Experimental Philosophy* (1794) included illustrations of a number of different devices used to explain the action of forces. Figures 1, 2, and 3 demonstrate different aspects of the law of the lever. They show various positions of equilibrium of a fulcrum and two weights or a weight plus an upward force exerted by a hand on a lever. In figure 5 (*lower left*), a force table demonstrates the parallelogram of forces rule (see Ill. 112) by showing how the weight on the right side is balanced by the two on the left. At the center, various combinations of pulleys are shown in equilibrium. The device at the far right is a wheel and axle, used for analyzing the turning moments of forces. A turning moment is measured by the product of a force and the radius of the circle through which it acts. The small weight acts through the radius of the disc to which it is attached, while the larger weight acts through the much smaller radius of the axle; the products of the weight and the radius are the same, so there is equilibrium and the wheel stays at rest instead of turning.

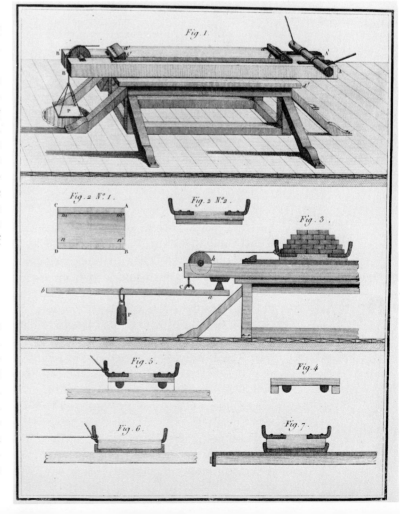

119. The experimental study of friction. The exact laws of friction were not known until Coulomb invented a way of studying the effects of friction under a variety of conditions of rest and motion. He used a small cart and a sledge, which were given different loads and were then subjected to forces of different magnitudes. Coulomb also made the important distinction between starting friction and moving friction. In addition, he studied the forces of torsion or twisting in rods and wires.

Fig. 1.   Fig. 2.

120. Huygens's analysis of relative motion. Christiaan Huygens, who first published the law of centrifugal force in 1673, also pioneered in the analysis of relative motion. In this "thought-experiment" (an experiment performed only in theory), a man in a boat and a man on the shore provide two different reference systems for viewing the movement of two balls suspended on strings, and thereby demonstrate the effect of relative motion. The man in the boat holds two balls of equal weight on strings of equal length. The two balls are made to strike each other with speed v, while the boat moves away from the man on shore. The observer in the boat (the uniformly moving frame of reference) will see the balls rebound with equal velocities (v), but the observer on land (the stationary frame of reference) will see one ball at rest and the other moving with the velocity (2v) obtained from the ball that has struck it. (The balls are supposed to be perfectly elastic.) In the first case, each moving ball gives to the other the velocity (and momentum) it had before impact; in the other, a moving ball gives a ball at rest all of its velocity (or momentum) and comes to rest. In this way Huygens began to analyze relative motions in relation to different frames of reference.

10

The Science of Motion

For centuries it had been an axiom of philosophers that to be ignorant of motion was to be ignorant of nature. Radical new ideas concerning motion were a conspicuous feature of the Scientific Revolution. There were three basic topics that were studied during the seventeenth century: uniform motion in relation to rest, the motion of freely falling bodies, and the motion that ensues when one body strikes another. The problem of impact proved to be especially important because it led to the first significant principle of conservation: the conservation of linear momentum, a major step in the development of the exact sciences. It also led to a concept of force based on impact, which was essential to the development of Newton's ideas about the causes of changes in motion.

The problem of freely falling bodies was solved by Galileo, who showed mathematically that—except for the effects of air resistance—bodies fall freely with a constant acceleration. He could not test this result by direct experiments with freely falling bodies,

so he invented the experiment of the inclined plane, in which he "diluted" gravity. He showed that the same kind of relationship exists between distance and time for a body moving along an inclined plane as that which he had found theoretically to hold for a freely falling body. Both are examples of motion with a constant acceleration, in which the distance is proportional to the square of the time. Galileo used this important discovery to analyze the paths of projectiles.

Making use of the law of composition of velocities, according to which the horizontal or forward motion of a projectile may be considered to be a component independent of the downward falling motion, Galileo proved that if a body moves forward at constant speed, while simultaneously falling freely with a uniformly accelerated motion, its path will be a parabola. Though many of his predecessors had developed fanciful theories about the motion of projectiles, it was Galileo who solved the puzzle. His solution was essential for our understanding of motion and was important in a practical way for the new, burgeoning science of exterior ballistics.

Another major problem in Galileo's day was to determine whether or not the Copernican system of the world could be tested by terrestrial experiment. For example, many people argued that if the earth were in motion, a weight dropped from a tower would not strike the ground directly below the tower, but would be "left behind" by the moving earth. They also supposed that if the earth were moving, a projectile fired vertically upward with a great velocity would not fall anywhere near the same spot from which it had been shot upward. A similar argument held that if two identical projectiles were fired, one due east and the other due west, the earth's movement would cause them to travel unequal distances.

Galileo explained that a weight dropped from the top of a mast on a moving ship would reach the deck at the foot of the mast, just as it would if the ship were at rest, save for the slight effects of air resistance. (Such an experiment, proving Galileo's contention, was carried out successfully in 1640 in Marseilles by Pierre Gassendi.) The concept implicit in Galileo's explanation was stated rigorously and generalized by Descartes, who declared that uniform motion could be described by the term "state," which had previously been reserved for the condition of rest. According to Descartes, if there is no external force acting on a body in uniform motion along a straight line, that body will continue in that motion unless it experiences a resistance, encounters some object, or is deflected by some force. This is true only for a rectilinear motion. This general principle has become known as the law of inertia.

It follows that if only linear motion may continue without an external force, then every curved motion requires the continuous action of an external force. In the seventeenth century, such a force was called centrifugal, because a body moving in a circular path appears to have a tendency to flee from the center. Actually it tends to move out along a tangent to its circular path and thus seems to be urged away from the center. It was not until Newton published his *Principia* in 1687 that the term "centrifugal force" was replaced by the more correct term, "centripetal force." Indeed, there was considerable confusion in the early seventeenth century as to whether or not circular motion may require a kind of equilibrium or balance between centripetal and centrifugal forces. Basing his analysis on the law of inertia as set forth by Descartes and on the nature of centripetal force, Newton was able to conclude that there exists a force of universal gravitation, which holds the solar system together and keeps the planets moving according to Kepler's three laws of planetary motion.

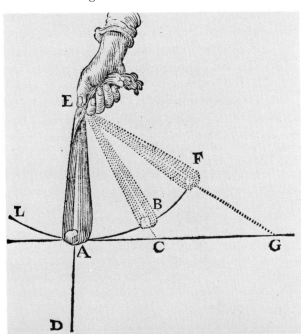

121. Centrifugal force. In his *Principia philosophiae* (1644) Descartes analyzed the forces that act when an object (in this case a ball in a sling) is made to move with uniform circular motion. Today, we would explain the tendency of the ball to move out from the center of the orbit as an effect of its inertia, which would actually tend to make it move out along the straight line *ACG*, taking it further and further away from the center of the circle. But seventeenth-century scientists who followed Descartes explained that this phenomenon arises from a "centrifugal force" of the revolving body that causes it to flee from the center and move out to an orbit of larger radius. Descartes was the first person to set forth clearly and distinctly the principle of linear inertial motion.

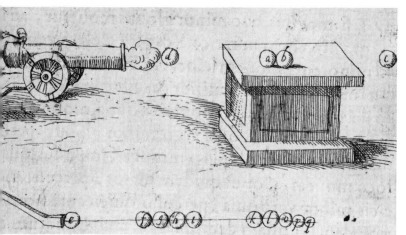

122. The laws of impact. Johannes Marcus Marci's *On the Proportion of Motion* (1639) includes this example of elastic impact. All the motion of a cannonball fired at a line of similar, touching cannonballs is transferred to the last ball in the line. The cannonballs are considered to be perfectly elastic, so that what the illustration depicts is the principle of conservation of momentum.

123. Glancing impact. Marci's book also dealt with the type of impact that occurs when one body receives a glancing blow from another, a frequent occurrence in billiards. (Note the fascinating mallet-shaped cues that were used.) The motions after impact illustrate the principles of conservation, analyzed in terms of the composition of velocity (or momentum).

124. Descartes's theory of diminishing weight. In a letter of April 1634 to the physicist and friar Marin Mersenne, Descartes referred to an experiment in which a large cannon, in the middle of a plain, was pointed directly upward to the zenith; when the cannon was fired, the ball did not fall back to earth again. Descartes wrote, "I would not judge it to be impossible," but he wanted to have the experiment tried. In another letter (15 May 1634), Descartes thanked Mersenne for having performed the experiment with an "arquebuse," but added that he could not consider the results to be certain unless a large cannon was used. In a third letter (13 July 1638), Descartes discussed some specific experiments and observations concerning apparent weight in relation to height above the earth. One example used was that "large birds such as cranes . . . have greater ease of flying high up in the air than lower down"; he judged that this effect results not so much from winds as that "the removal of the birds from the earth makes them lighter." If one could believe Mersenne's experiment with the musket and the reports of others with cannons, one would have to conclude that the force of the cannon carried the balls "so far from the earth that this caused them to lose all of their weight," and they would never fall to earth again. This cartoon shows a priest (Mersenne) and a philosopher (Descartes) performing the experiment; the banner reads, "Will it fall down again?"

125. Pre-Galilean diagram of the path of a projectile. Sebastian Münster's *Rudiments of Mathematics* (1551) shows the trajectory of a projectile as explained by Tartaglia. The trajectory consists of a straight line out from the mortar or cannon, a small circular arc, and then a straight line downward.

126. The parabolic path of projectiles. Excluding any effects of air friction, projectiles have parabolic paths. This was one of Galileo's great discoveries. The parabolic paths were illustrated in a book on experimental and mathematical physics published by Petrus van Musschenbroek in 1762. The upper figure depicts a series of parabolic paths for a horizontal projection at different velocities; the lower figure delineates projections at different angles to the horizontal.

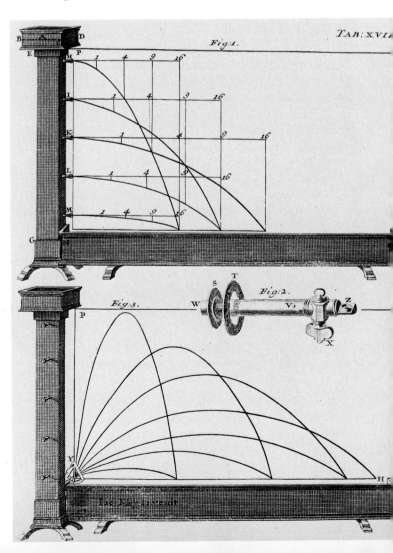

11

ℕewtonian Science

The Scientific Revolution is generally agreed to have reached its full state of maturity with the advent of Newton's *Principia* in 1687. In this great book, Newton set forth the principles of dynamics and expounded his system of the world, based upon his concept of universal gravitation. He showed that a force of gravity acts between any two bodies wherever they may be in the universe. The magnitude of this force is directly proportional to the product of the masses of the two bodies and inversely proportional to the square of the distance between them.

Newton also invented the calculus, which has become the language of dynamics, but in his *Principia* he did not consistently apply this new mathematical method to the physical problems discussed. Possibly, he did not want to ask his readers to learn both a new kind of mathematics and a new kind of physics, the latter being difficult enough. Furthermore, he took pride in presenting his work in the traditional language of geometry and proportion, thus firmly establishing his lineage with

the great geometers of Greek antiquity.

Newton systematized the laws of motion into three axioms, which he used to elucidate the physical bases of Kepler's three laws of planetary motions. First, Newton showed that Kepler's first law—that a line from a center to a moving body (such as a planet) sweeps out equal areas in equal times—is merely a necessary and sufficient condition for the existence of a centrally directed or centripetal force. Kepler's second law—that orbits are ellipses (or, in special cases, circles)—was shown to imply that the centrally directed force varies inversely as the square of the distance from the center. Kepler's third or harmonic law was shown by Newton to imply that one and the same force acts on all the planets.

In the science of the *Principia*, the force of universal gravity does more than determine the motions of planets in Keplerian orbits. Newton made use of the concept to explain the phenomena of the tides, the precession of the earth's axis, and the falling of bodies ac-

127. Isaac Newton. This portrait by Godfrey Kneller was made when Newton was at the height of his creative powers. It shows his long, sensitive face and the artistic hands with which he designed and performed experiments of great delicacy and accuracy.

cording to Galileo's discoveries—even to predict the shape of the earth as an oblate spheroid. Above all Newton was able to show that one simple law could account for many diverse phenomena of the earth and the heavens. He gave terrestrial and celestial science a unity that had never existed before, except in men's hopes.

Newton's great triumph produced both admiration and opposition. Some opponents were Cartesians, reluctant to change their intellectual ways. Others were scientists and philosophers who objected on general principles to the concept of a grasping force that could spread out from the sun to Saturn, and even beyond Saturn to the end of the orbits of comets, and still affect the motions of heavenly bodies. The introduction of such a force seemed to be a return to "occult" notions, which the "mechanical philosophy" had supposedly banished from science.

Newton himself found difficulty with the concept of "action at a distance," and proposed various explanations (including the action of an all-pervading ether) to explain how gravitation might occur. His ultimate point of view with respect to this problem was expressed in the concluding General Scholium of the *Principia*: Even if we cannot understand how the force of gravitation works, "it is enough" (*satis est*) that we can know the laws by which this force operates, and that by means of this force we can accurately predict and retrodict the phenomena of the external world.

In the eighteenth century, the Newtonian natural philosophy and system of the world became generally accepted. Scientists in other fields came to look upon Newton's work as a model, a goal toward which all the sciences should strive. Thinkers in fields other than the exact sciences also expressed the hope that their studies of man and society might reach the same high level that Newton had achieved in his *Mathematical Principles of Natural Philosophy*. More than anyone else, Newton symbolized the enormous achievements of which human reason is capable, once its powers are enlarged by true science and disciplined by mathematics. He also made significant contributions to the fields of mathematics and optics (Chapters 8 and 12).

128. Newton's early thoughts on motion. During an early stage of his intellectual development, Newton kept a *Waste Book* of his thoughts on various aspects of science. The first definition at the top of this page says that "When a body passeth from one parte of Extension to another it is saide to move." In the middle of the page, he discusses refraction, coupled with a drawing of a refracted ray of light, and the definition: "Refraction is when the body *c* passing obliquely through the surface *ed* at the point *b* meets with more opposition on one side of the surface than on the other & soe looseth its Determinacon; as if it turne towards *a*." A series of "Axiomes & Propositions" follows at the bottom of the page, starting with two that declare the principle of inertia, which Newton learned about by reading Descartes's *Principia*. Newton later combined them to derive his first law of motion. But here he says: "1. If a quantity once move it will never rest unlesse hindered by some externall caus. 2. A quantity will always move on in the same streight [line] (not changing the determination nor celerity of its motion) unlesse some externall cause divert it."

AXIOMATA,
SIVE
LEGES MOTUS.

LEX I.

Corpus omne perseverare in statu suo quiescendi vel movendi uniformiter in directum, nisi quatenus a viribus impressis cogitur statum illum mutare.

Projectilia perseverant in motibus suis, nisi quatenus a resistentia aeris retardantur, & vi gravitatis impelluntur deorsum. Trochus, cujus partes cohaerendo perpetuo retrahunt sese a motibus rectilineis, non cessat rotari, nisi quatenus ab aere retardatur. Majora autem Planetarum & Cometarum corpora motus suos & progressivos & circulares in spatiis minus resistentibus factos conservant diutius.

LEX II.

Mutationem motus proportionalem esse vi motrici impressae, & fieri secundum lineam rectam qua vis illa imprimitur.

Si vis aliqua motum quemvis generet; dupla duplum, tripla triplum generabit, sive simul & semel, sive gradatim & successive impressa fuerit. Et hic motus (quoniam in eandem semper plagam eum vi generatrice determinatur) si corpus antea movebatur, motui ejus vel conspiranti additur, vel contrario subducitur, vel obliquo oblique adjicitur, & cum eo secundum utriusque determinationem componitur.

LEX

LEX III.

Actioni contrariam semper & aequalem esse reactionem: sive corporum duorum actiones in se mutuo semper esse aequales & in partes contrarias dirigi.

Quicquid premit vel trahit alterum, tantundem ab eo premitur vel trahitur. Si quis lapidem digito premit, premitur & hujus digitus a lapide. Si equus lapidem funi alligatum trahit, retrahetur etiam & equus (ut ita dicam) aequaliter in lapidem: nam funis utrinque distentus eodem relaxandi se conatu urgebit equum versus lapidem, ac lapidem versus equum; tantumque impediet progressum unius quantum promovet progressum alterius. Si corpus aliquod in corpus aliud impingens, motum ejus vi sua quomodocunque mutaverit, idem quoque vicissim in motu proprio eandem mutationem in partem contrariam vi alterius (ob aequalitatem pressionis mutuae) subibit. His actionibus aequales fiunt mutationes, non velocitatum, sed motuum; scilicet in corporibus non aliunde impeditis. Mutationes enim velocitatum, in contrarias itidem partes factae, quia motus aequaliter mutantur, sunt corporibus reciproce proportionales. Obtinet etiam haec Lex in Attractionibus, ut in Scholio proximo probabitur.

COROLLARIUM I.

Corpus viribus conjunctis diagonalem parallelogrammi eodem tempore describere, quo latera separatis.

Si corpus dato tempore, vi sola M in loco A impressa, ferretur uniformi cum motu ab A ad B; & vi sola N in eodem loco impressa, ferretur ab A ad C: compleatur parallelogrammum $ABDC$, & vi utraque feretur id eodem tempore in diagonali ab A ad D. Nam quoniam vis 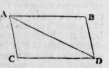 N agit secundum lineam AC ipsi BD parallelam, haec vis per Legem II nihil mutabit velocitatem accedendi ad lineam illam BD a vi altera genitam. Accedet igitur corpus eodem tempore ad lineam BD, sive vis N imprimatur, sive non; atque adeo in fine illius temporis reperietur alicubi in linea illa BD. Eodem argumento in fine temporis ejusdem reperietur alicubi in linea CD, & idcirco in utriusque lineae concursu D reperiri necesse est. Perget autem motu rectilineo ab A ad D per Legem I.

B 3   COROL-

---

[402]
REGULAE PHILOSOPHANDI.
HYPOTHESES.

Reg.
Hypoth. I. *Causas rerum naturalium non plures admitti debere, quàm quae & verae sint & earum Phaenomenis explicandis sufficiant.*
Natura enim simplex est & rerum causis superfluis non luxuriat.

Reg. II.
Hypoth. II. *Ideoque effectuum naturalium ejusdem generis eaedem sunt causae.*
Uti respirationis in Homine & in Bestia; descensus lapidum in Europa & in America; Lucis in Igne culinari & in Sole; reflexionis lucis in Terra & in Planetis.

Hypoth. III. *Corpus omne in alterius cujuscunque generis corpus transformari posse, & qualitatum gradus omnes intermedios successive induere.*

Hypoth. IV. *Centrum Systematis Mundani quiescere.*
Hoc ab omnibus concessum est, dum aliqui Terram alii Solem in centro quiescere contendant. PHAENOMENA.

Phaenom. I.
Hypoth. VI. *Planetas circumjoviales, radiis ad centrum Jovis ductis, areas describere temporibus proportionales, eorumque tempora periodica esse in ratione sesquialtera distantiarum ab ipsius centro.*
Constat ex observationibus Astronomicis. Orbes horum Planetarum non differunt sensibiliter à circulis Jovi concentricis, & motus eorum in his circulis uniformes deprehenduntur. Tempora verò periodica esse in ratione sesquialtera semidiametrorum orbium consentiunt Astronomi: & Flamstedius, qui omnia Micrometro & per Eclipses Satellitum accuratius definivit, literis ad me datis, quintuplici numeris suis mecum communicatis, significavit rationem illam sesquialteram tam accurate obtinere, quàm sit possibile sensu deprehendere. Id quod ex Tabula sequente manifestum est.

Satellitum

---

129. Newton's three laws of motion. Here are the three laws of motion, stated to be "Axioms or laws of motion," as they were printed in the first edition of Newton's *Principia* (1687). Law one is a formal statement of the Cartesian principle of inertia. Law two presents the proportionality of an external force and the change in momentum it produces. Law three declares that "Reaction is always equal and opposite to action." Corollary 1 to the laws sets forth the parallelogram law for combining motions. Since this corollary supposes that a blow or impulsive force gives a body motion, it shows that the second law states a relation between impulsive forces and the motions they produce.

130. Newton's "hypotheses." Although Newton's natural philosophy is usually characterized by his slogan "I do not frame [feign] hypotheses" (*Hypotheses non fingo*), this statement was not present in the first edition of the *Principia*. In fact, Newton originally began Book Three, "On the System of the World," with a series of hypotheses. A page from Newton's personal copy of the first edition shows how he altered the hypotheses. He has crossed out the general heading "HYPOTHESES" and has replaced it by two new ones: "REGULAE PHILOSOPHANDI" (or "Rules of Proceeding in Natural Philosophy") at the top of the page, and "PHAENOMENA" in the middle of the page. The first two hypotheses have become "Reg. I" and "Reg. II," while those numbered three and four have been crossed out. The old "Hypoth. V" has become "Phaenom. I."

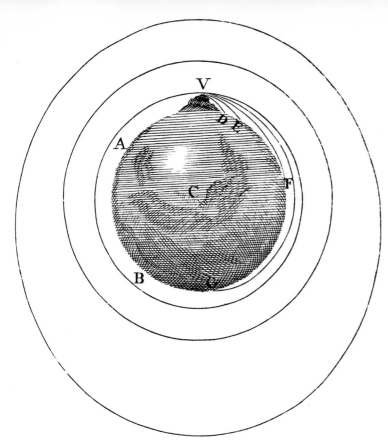

131. The first artificial satellite. In his popular exposition *System of the World* (first published posthumously in 1728 in Latin and English), Newton depicted the results of firing a projectile from a high mountain at several different velocities. Ordinarily, a projectile would strike the earth at distances increasing in relation to greater muzzle velocity. But Newton argued that if the velocity was made sufficiently great, a projectile would not strike the earth at all but would move about it in orbit.

132. The path of least descent. In 1696 Johann Bernoulli published a challenge problem for all mathematicians of the world: To find the path along which a body would fall in the shortest time, if the path is not a vertical straight line. He postulated that the path of swiftest descent would turn out to be curved, even though a straight line is the shortest distance from one point to another. Newton received a copy of the challenge from Bernoulli and promptly solved it that same evening, after completing a full day's work at the mint where he was then employed. He published the solution anonymously. When Bernoulli read it, however, he recognized at once that only Newton could have found the solution in so elegant a manner, and he exclaimed, *"Ex ungue, leonem"* ("The lion is known by his claw"). The solution to the problem is a curve known as the brachystrochrone, which is a cycloid. Some eighteenth century, devices, such as the one pictured here, demonstrated the solution to the problem by means of two tracks, one straight, the other a cycloid. If two balls are simultaneously released at equal heights, one along each track, starting at the upper right, the ball rolling along the cycloidal curve will reach the point of intersection before the one on the straight path. The greater steepness of the curve in its initial part makes the ball accelerate faster than in the corresponding part of the straight line.

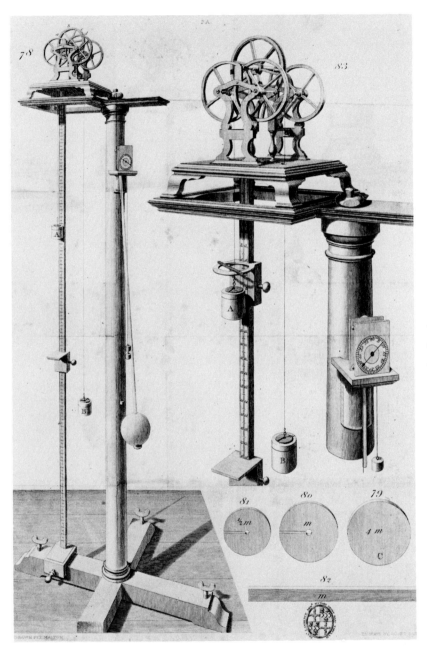

133. Atwood's machine for demonstrating Newton's second law of motion. This apparatus, described in George Atwood's *Treatise on the Rectilinear Motion and Rotation of Bodies* (1748), uses a pair of weights connected by a string that passes over a series of geared wheels. It is designed to test Newton's second law, namely, that the acceleration produced on a given mass is directly proportional to the accelerative force. A tiny notched weight or "rider," such as the one marked 80, or a small bar, such as 82, could be added to or taken away from one of the moving pairs of weights, whose rate of descent had already been measured. Since the motion of the weights is much slower than the motion of free fall, exact measurements of the differences in velocity (or acceleration) of a body can be made when the small weights are added or subtracted. In every case, the acceleration proves to be proportional to the force, as Newton had predicted.

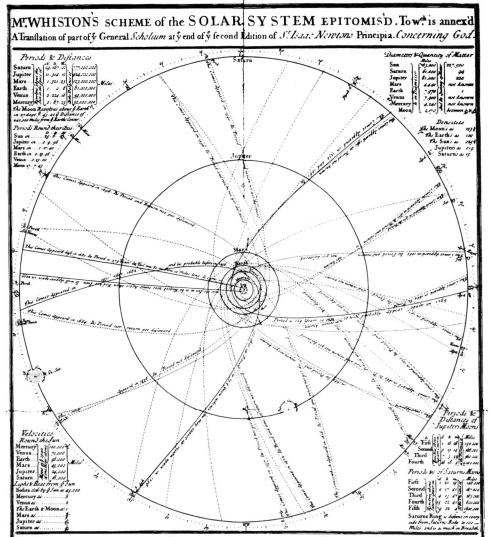

134. Whiston's scheme of the solar system. In 1724 William Whiston, Newton's successor as Lucasian professor at Cambridge University, published the broadside "Scheme of the Solar System Epitomis'd." To it he subjoined a translation of a portion of the General Scholium with which Newton's *Principia* concludes, and which had been published for the first time in the second edition of the *Principia* in 1713. In the diagram the planets, surrounded by their satellites, are moving in empty space around the sun under the action of the force of gravity. Also included are the orbits of comets moving through empty space under the action at the same gravitational force.

135. A memorial to Newton in the 1720s. Commissioned in Italy by Owen MacSwinny, an Irish theatrical manager, as one of a series of two dozen paintings, all of illustrious Britons, this painting (like the others) is dominated by a large "Urn, wherein is supposed to be deposited the Remains of the deceased Hero." Begun after Newton's death in 1727, it was finished by 1730. There are many symbols of science in the painting, including a globe and drawing or measuring instruments at lower right, an armillary sphere at the foot of the pedestal supporting the urn, and some pages of mathematics at right center. The only truly Newtonian element in the design is the representation of a prism experiment. The artist, however, has not shown the actual experiment, but has merely displayed the formation of a spectrum by a prism. The painting was primarily the work of Giovanni Battista Pittoni.

Part Four

THE
EXPERIMENTAL
PHYSICAL
SCIENCES

VITELLIONIS MA-
THEMATICI DOCTISSIMI περὶ ὀπτικῆς,
id eſt de natura, ratione, & proiectione radiorum uiſus, luminum, colorum atcʒ formarum, quam uul-
go Perſpectiuam uocant,
LIBRI   X.

Habes in hoc opere, Candide Lector, quum magnum numerum Geometricorum
elementorum, quæ in Euclide nuſquä extant, tum uero de proiectione, infractione, &
refractione radioʒ uiſus, luminum, colorum, & formarum, in corporibus transparenti-
bus atcʒ ſpeculis, planis, ſphæricis, columnaribus, pyramidalibus, cöcauis & conuexis,
ſcilicet cur quædam imagines rerum uiſarü æquales, quædä maiores, quædam minores,
quædam rectas, quædä inuerſas, quædam intra, quædä uero extra ſe in aëre magno mi-
raculo pendentes: quædam motum rei uerum, quædä eundem in contrariü oſtendant:
quædä Soli oppoſita, uehementiſſime adurant, ignem'cʒ admota materia excitent: de'cʒ
umbris, ac uarijs circa uiſum deceptionibus, à quibus magna pars Magiæ naturalis de-
pendet, Omnia ab hoc Autore (qui eruditorum omniü conſenſu, primas in hoc ſcripti
genere tenet) diligentiſſime tradita, ad ſolidam abſtruſarum rerum cognitionem, non
minus utilia cʒ iucunda. Nunc primum opera Mathematicoʒ præſtantiſſ. dd. Ge-
orgij Tanſtetter & Petri Apiani in lucem ædita.

Norimbergæ apud Io. Petreium, Anno MDXXXV.

136. The classical problems of optics. An illustration from the writings of the
thirteenth-century Polish physicist Witelo (published in 1535) pictures some of the
traditional problems relating to light that were studied by optical theorists. One
is the transmission of light in straight lines, represented by a ray extending from
the sun. Another is the calorific power of the sun's rays, producing fire when the
sun's rays are focused on an object by means of a lens. Reflection is illustrated
by a mirror, and refraction by the bent image of a man's legs partially submerged
in water. And, of course, there is a rainbow.

12

Optics

Optics is a scientific subject that can be traced back at least as far as Euclid, who wrote a special work on the subject, as did Ptolemy. Those who studied optics had always sought to explain how light is reflected and how it is refracted, or bent, as it goes from one medium into another, for example, from air into water. Interest was also directed to finding out why light travels in straight lines, and there was much debate about whether we see an object by means of light beams coming to the eye from an illuminated object or by means of visual rays emanating from the eye in the direction of the object seen. Vestiges of the latter theory are enshrined in such verbs associated with seeing, as "to peer," or to "cast a glance." Another phenomenon that long defied understanding was the formation of rainbows.

In the Renaissance, there were two major innovations in the study of optics. One was the introduction of eyeglasses, which raised the problem of explaining the action of negative (biconcave) lenses, as well as positive (bi-

convex) lenses. (Traditional accounts of the discovery of the telescope attribute the new instrument to chance, a chance combination —unguided by optical theory—of positive and negative lenses for eyeglasses.) The other innovation was the invention, or reinvention, of the science of perspective.

Of great significance was Kepler's discovery that the eye is a *camera obscura*, or dark chamber, in which the lens causes the image of an external object to be focused on the retina. The concept of a lens system in the eye was intimately related to seventeenth-century discoveries about the action of lenses and the refraction of light.

The law of refraction, known today as Snel's Law, was discovered by Willebrord Snel in Holland, who did not publish his discovery. The law was made available in 1637 by Descartes, who may have found a clue in Snel's manuscript. In fact, Snel's countrymen charged Descartes with plagiarism.

Two major explanations of optical phenomena were set forth in the seventeenth century: the wave theory, whose chief advocates were Hooke and Huygens; and the corpuscular theory, whose chief advocate was Newton. The wave theory attributed all optical phenomena to the existence of trains of waves in an all-pervading ether, while the corpuscular theory assumed the phenomena to arise from streams of very tiny bodies emanating from sources of illumination. A major argument favoring the corpuscular theory was the rectilinear propagation of light and the fact that light does not seem to bend or spread into the shadows as sound does. But Newton discovered a whole class of phenomena (now known as interference and diffraction) that indicated that periodicity is associated with light and periodicity implies some kind of wave phenomena.

Newton also advanced our understanding of color through a famous series of experiments in which he demonstrated that sunlight is a mixture of light of different colors, each of which has its own index of refraction, or is bent to a different degree whenever light travels from one medium to another. He also devised an experiment to show that, unlike sunlight, monochromatic light does not break down into components; it is "simple" rather than "compound."

Newton's studies on color and refraction led him to believe that all images produced by lenses must suffer from what we call chromatic aberration, a phenomenon in which light of different colors is brought to a focus at different points by a simple lens. He thereupon invented the reflecting telescope, in which light is brought to a focus not by being refracted through a lens, but by being reflected from the surface of a magnifying mirror.

137. Eyeglasses of the seventeenth century. These forms of eyeglasses were illustrated in an Italian book of 1689. Of particular interest are the glasses worn by the man on the left, which consist of dark screens with a tiny hole in the center.

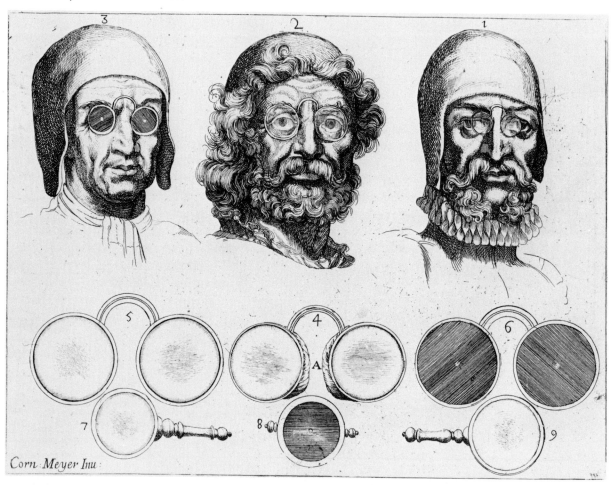

Corn: Meyer Inu:

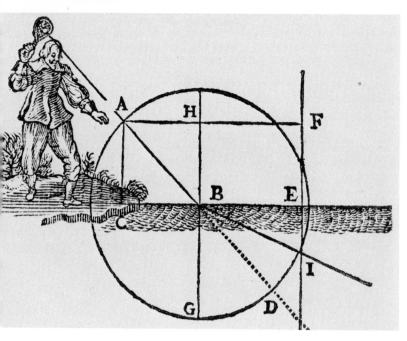

138. The law of refraction. The true law of refraction discovered by W. Snel, a Dutchman, was first made known in Descartes's book *Dioptrics* (1637), to which his famous *Discourse on Method* was a preface. Descartes compared the motion of light to that of a tennis ball driven through the boundary, or interface, between the air and a pool of water. The path of a tennis ball would turn up toward the surface of the water (*BI*), but Descartes showed how the path of light is directed down, away from the surface (not shown). He found the quantitative form of the law of refraction and thus solved a problem that had vexed investigators of physical science since the days of Ptolemy in the second century A.D., if not longer.

139. Formation of the primary and secondary rainbows. Descartes published an analysis of rainbows in his *Dioptrics*. This illustration, taken from an eighteenth-century book on general physics, shows Descartes's analysis of how the primary and secondary bows are formed. Light coming from the sun enters the top of raindrops (where it is refracted), is reflected internally, and then exits from the drop (in a second refraction), and travels in a straight line to the observer's eyes, producing a primary rainbow. Descartes used his knowledge of the laws of reflection and refraction plus his skill as a mathematician to show by computation why the angle formed by the sun's rays and the rays from the bow to the observer's eye (*EOF*) is always about 42°. A secondary rainbow may be formed when sunlight enters the bottom of a drop and is reflected twice before leaving the upper part of the drop to reach the eye of the observer. Descartes showed why the angle for the secondary bow is about 52°. Newton's discovery of the law of separation of sunlight into colors explained the formation of colors in the rainbow, and provided the reason why the emerging colors are separated at a constant set of angles that depend on the index of refraction in light and in air. Thus, Newton completed the work of Descartes by explaining why the order of the colors in the primary bow is the reverse of those in the secondary bow. He also showed why these bows have the widths that we observe.

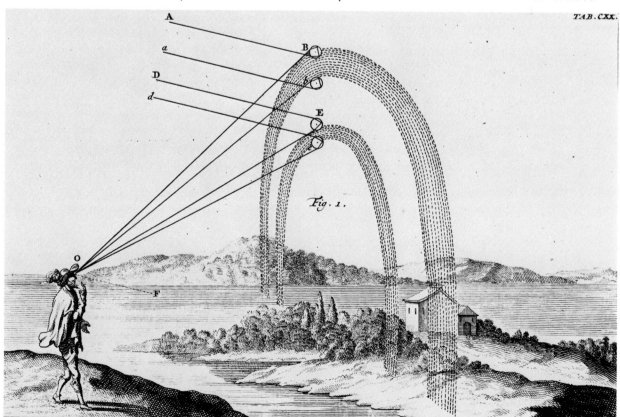

140. Newton's sketch of his experiment on light and color. In this classic experiment, light from the sun enters a chamber through a small hole in a window shutter (*right*). It then may pass through either a lens alone or through a lens plus a prism. Light that goes through the lens alone continues in a straight line and comes to a focus at a single spot on the lower portion of the board at left. Light that passes through the prism is separated into various colors that form a spectrum on the upper half of the board. With this apparatus, Newton demonstrated that sunlight, or white light, is a mixture of different colors, each having its own degree of refraction when passed through a prism. By drilling small holes far enough apart in the board so that only a single color of light could pass through each one, and then placing a second prism behind the board, he was also able to show that light of a particular color is refracted as it passes through the second prism, but its color does not change. In the upper left hand corner Newton has written, *"Nec variat lux fracta colorem":* "Nor does the refracted light change color [on a second refraction]," that is, "Light is not further broken down into component colors."

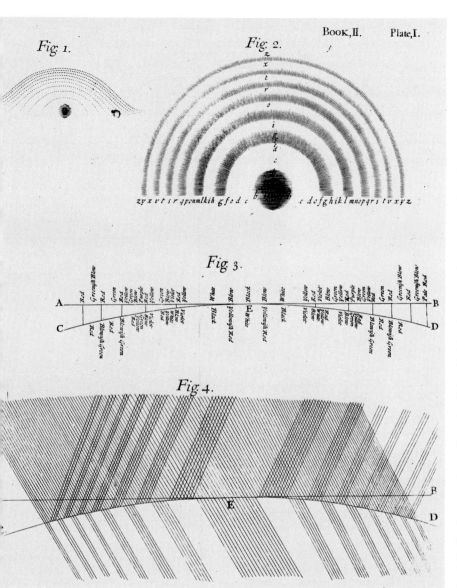

Fig. 1.   Fig. 2.

Fig. 3.

Fig. 4.

141. "Newton's rings." This diagram from Newton's *Opticks* (1704) illustrates the phenomenon of rings produced when light passes through a circular, wedge-shaped, narrow space. Figure 3 illustrates such a space in cross-section; it lies between a flat surface of glass and the convex surface of a lens. The appearance of the rings is shown at the top (Figure 2) and the space in which they are produced is at the center (Figure 3). The top plane *AB* marks one boundary of the region of air that produces the rings, and the curved surface *CD*, representing the upper convex face of the lens, marks the lower boundary. Figure 4 illustrates Newton's explanation of the successive reflection and transmission of light that produces the rings. In one experiment, shown in Figure 1, Newton pressed two glass prisms tightly together, but owing to slight imperfections in their surfaces, small pockets of air were trapped between them. When he turned the prisms so they were, in his words, slightly "inclined to the incident rays of light," Newton produced a set of "slender Arcs of Colours" around a central transparent spot. By turning the prisms further, the arcs increased in length and eventually became full circles or rings, a classic example of an interference phenomenon.

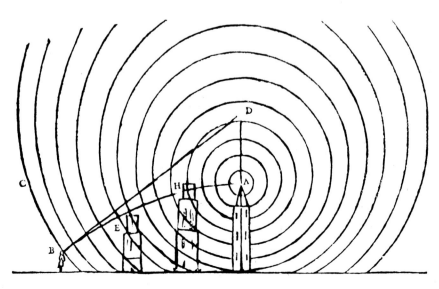

142. Huygens's explanation of atmospheric refraction. Christiaan Huygens espoused a wave theory of light in a treatise published late in life in 1690. He conceived that light is propagated from a luminous source in a series of waves or undulations in an all-pervading ether. In this diagram, light from the top of a tower (*A*) travels to the eye of an observer (*B*) in a series of waves that spread out in all directions. When the wave *C* reaches the observer, he will see the top of the tower, but the refraction of the air will make it seem to him that he is looking in the direction *BD*.

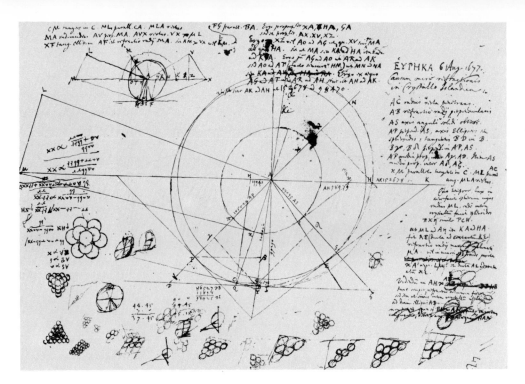

143. Huygens's explanation of the double refraction in Iceland spar. One of the great achievements of Huygens was to explain, by means of his concept of wave envelopes, how certain crystalline substances, such as calcite, produce two different rays by refraction: an ordinary ray and an extraordinary ray. Huygens's solution was to envisage that in calcite there are two types of waves of light, the spherical (corresponding to the ordinary ray) and the spheroidal (corresponding to the extraordinary ray). Since each type of wave or ray produces an image, any object seen through the crystal will appear double. On this page of his original manuscript Huygens has sought the cause of this phenomenon: the arrangement of the particles that constitute the crystal is regular (*center* and *upper left*). In the upper right-hand corner, Huygens wrote "Eureka," echoing Archimedes's famous expression "Eureka—I have found it" on solving the problem of the amount of gold alloyed in Hiero's crown. The whole statement reads: "Eureka 6 Aug. 1677, the cause of the wonderful refraction of Iceland Spar."

144. Photometry. One of the great advances in the science of light in the century following Newton and Huygens was the discovery of a means of measuring intensities of light. This diagram explains the use of the photometer invented by Johann Heinrich Lambert. It was described in his treatise *Photometria* (1760), which was devoted to the measurement and gradation of light, color, and shadows. The illustration demonstrates how Lambert compared the intensities of two light sources, such as candles. He placed one in the foreground at *A* and the other in the rear at *K* and set up an opaque screen (*HI*) in front of a large, white, smooth wall (*EGBC*). The light sources were arranged so that *A* would produce a shadow of the screen that would cover the far half of the wall (*DCEF*), while the shadow cast by *K* would fill the near portion of the wall (*BDFG*). Thus, each half of the wall would be illuminated by one of the two sources. To obtain different values of light intensity, one of the sources (*A*) would then be moved back and forth until the wall appeared equally bright on either side of the dividing line *DF*. Lambert made a number of significant photometric measurements, among them the ratio of the mean brightness of the full moon to that of the sun (1:277,000).

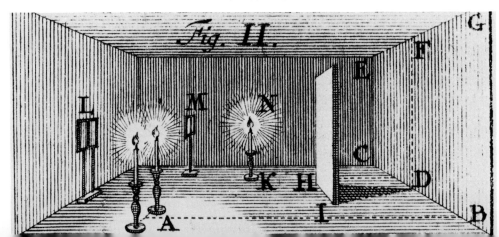

13

The Microscope

When we think of the microscope today, we usually have in mind the compound microscope, an instrument with two lenses (or lens systems)—the eyepiece or ocular and the objective. Soon after the spectacular use of the telescope by Galileo, a number of scientists conceived of using a similar two-lensed instrument for studying the minute details of living things. In the seventeenth and eighteenth centuries, microscopes were developed with great magnification, capable of revealing fine details that had been hidden from view until that time.

Unfortunately, as the magnifying power of the compound microscope increased, the actual resolution decreased. Under high magnification, fine details would very often be blurred rather than shown clearly. Accordingly, much of the best work in microscopy in the seventeenth and eighteenth centuries was done with a simple microscope, a single lens of great magnification. Toward the end of the eighteenth century, the compound microscope was improved by introducing achromatic lens

combinations that reduced chromatic aberration; but it was not until the nineteenth century that the apochromatic microscope was invented, finally giving microscopists the high-powered tool with good resolution that had been so long desired.

Many important discoveries were made with the microscope in the last decades of the seventeenth century. One of the major achievements was to trace out the capillaries that link the arterial and venous systems, thus completing or proving William Harvey's doctrine of the circulation of the blood.

The most astonishing of the discoveries made with the microscope was probably the existence of the spermatozoa, first seen by the Dutch microscopists, of whom the most renowned was Antoni van Leeuwenhoek. Leeuwenhoek himself had already discovered many "very little animalcules," as he called them, or tiny forms of life (which later proved to be minute plants as well as minute animals). In addition to pioneering the observation of protozoa and bacteria, Leeuwenhoek discovered

145. Hooke's compound microscope. This microscope is said to have belonged to Robert Hooke, and to have been made according to his specifications by Christopher Cock. Hooke described the microscope and included a diagram of its construction in a preface to his *Micrographia* (1665). The instrument was a true compound microscope, with two lenses—an eyepiece at the top and an objective at the bottom. It could be used for studying transparent material in slides or for examining opaque materials.

parthenogenesis (in aphids) and budding in an animal, and did noteworthy research on blood corpuscles and the capillaries. Leeuwenhoek's discoveries in the microscopic world of plants and animals were so startling that at first they were not fully believed. Leeuwenhoek ground tiny lenses by hand and fixed them between a pair of perforated plates. He ground about five hundred such lenses, some of which were of 500 power, and which were not surpassed in quality until the nineteenth century.

The compound and simple microscopes of the seventeenth and eighteenth centuries could usually be used in two different ways. One was for the study of thin sections of plant or animal material, as in the microscope slides with which we are familiar today. In this case, the light, either sunlight or artificial light, was usually reflected through the speci-men into the objective lens of the microscope by means of a substage mirror. The second use was to study opaque objects, such as bits of coral, gems, and so on. In this case, the light had to come from above or from the side, since the specimen under examination would not transmit light.

In the eighteenth century, microscopes were also used to project large images that could be seen simultaneously by a group of viewers. The most popular form, called a "solar microscope," could be attached to a shutter of a window. Light from the sun would be reflected by an outside mirror so as to pass through the microscope and throw a large image on a screen or on a wall on the opposite side of a darkened room. Another form of projection microscope used artificial light and projected the image on a screen at the large end of a pyramidal box.

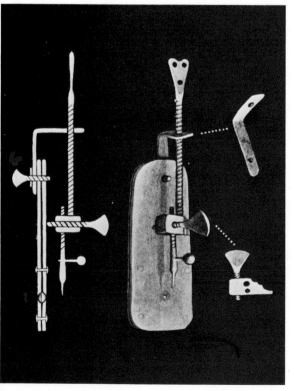

146. Leeuwenhoek's microscope. This drawing by historian-microscopist Clifford Dobell clearly delineates the structure of the instrument with which Leeuwenhoek revolutionized biology in the second half of the seventeenth century. Leeuwenhoek's precision and extraordinary dexterity permitted him to create lenses capable of magnifying objects 500 times. Since he was working with resolutions as fine as one micron (one millionth of a meter), it was necessary for him to devise new standards for measuring extremely small objects. Leeuwenhoek's microscope consisted of a simple magnifying lens, or single biconvex lens, mounted between two thin plates of brass or other metal. The specimen to be studied was placed on the tip of a small pointer set in front of the tiny lens (shown near the bottom of the left-hand figure, just to the left of the pointer tip). A transverse screw near the pointer (shown in detail in the lower right of the illustration) could be turned so as to bring the specimen closer to the lens or move it further away, thus serving as a focusing device. A long screw (shown vertically in the left-hand figure) was attached to a right-angled arm at the top and enabled the observer to move the specimen up or down across the lens, while a third screw provided a pivotal motion that permitted the specimen to be moved sideways across the lens. Leeuwenhoek would hold the instrument in front of his eye and look through the specimen against the sun or the light of a fire. He devised a slightly different arrangement for studying liquids and specimens in liquids.

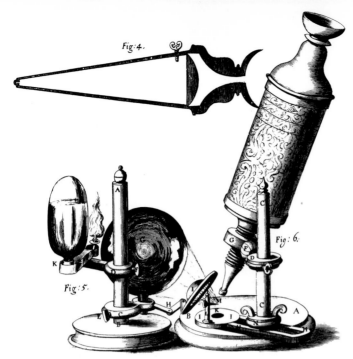

147. Studying an opaque object with Hooke's microscope. As seen in this diagram from Hooke's *Micrographia*, the microscope could be focused by rotating the whole instrument in a threaded collar (*G*) attached to a vertical stand. A cross section of the barrel turned on its side, at the top of the plate, shows a double-convex lens, or "object glass," of short focal length with a pinhole diaphragm at the left end of the barrel. The eyepiece, consisting of an "eyeglass" and "field lens," is surmounted by a kind of cup to ensure that the observer's eye is placed at the proper distance from the eyepiece. The field lens enlarges the field of view, so that the viewer can see more of an object at once. To examine smaller areas of an object, the observer would remove the field lens. The barrel of the microscope is made of nesting cardboard tubes, so that its length can be varied. Light from a spirit lamp passes through a condenser system consisting of a sphere, which concentrates the light that passes through a plano-convex lens onto the pointer (*M*) holding the specimen to be studied.

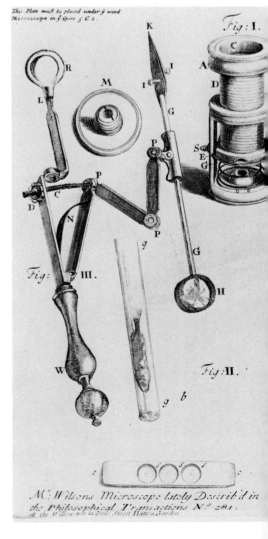

148. "Mr. Wilson's Sett of Pocket-Microscopes," or the screw-barrel microscope. This form of microscope, an improvement on existing instruments, was devised and manufactured by James Wilson in London and described by him in 1702. In Fig. I the microscope is delineated upside down, that is, it is standing vertically on the eyepiece or magnifying glass. It had nine lenses of different powers (*M*) that could be screwed into a mount. In use, a slide containing a specimen was inserted into a spring holder at the lower end (*E*) of the screw barrel (*D*); the barrel could then be turned, or screwed, in its mount to adjust the distance of the lens from the slide. A convex condensing lens (*C*) helped to provide even and concentrated illumination through the barrel. At the bottom of the page there is shown a slide with three openings for mounting transparent specimens between two thin glass plates. At the left (Fig. III) is another form of simple microscope. To use this one, a transparent specimen or a slide was clamped in place by a pair of spring jaws or tongs (*K*), or an opaque object was placed on a wooden platform (*H*). The lens (*M*) was screwed into the ring (*R*) at the end on an arm (*L*) that could be brought near the specimen. A screw and nut (*D,C*) moved the lens for focusing. This illustration, based on the account of the screw-barrel microscope published by Wilson in the *Philosophical Transactions of the Royal Society* (1702), comes from the supplementary volume of John Harris's dictionary *Lexicon technicum* (1710).

149. Lyonet's dissecting microscope. This type of instrument was used by Pierre Lyonet in making preparations for microscopy and for studying microscopic objects. Lyonet fixed a powerful single lens in a mount at the end of an arm made of linked "Musschenbroek nuts," which permitted movement of the lens to any position above the specimen. The specimen was set over a hole in a circular stand and illuminated by light reflected by the substage mirror. This particular microscope was developed in the early eighteenth century.

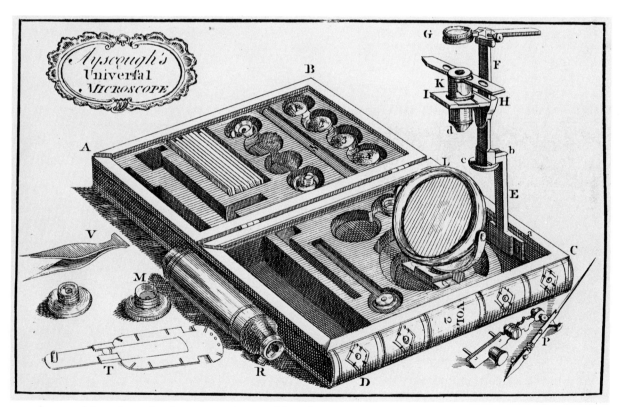

150. Ayscough's universal microscope. A portable microscope, made and sold by James Ayscough, was described in the *Universal Magazine of Knowledge and Pleasure* for February 1750. It could be packed away in a book-shaped box and stored on the shelf of a library. In this illustration a slide is being viewed through a simple lens, but the microscope also had a barrel with a compound lens system that is seen lying on the table beside the box. Because it could be either a simple or a compound instrument, the microscope was "universal." It could be used with a reflecting mirror, as shown, but also had devices for studying opaque objects or liquid specimens.

151. Mother and daughter using a microscope. This illustration is taken from the Abbé Jean-Antoine Nollet's popular five-volume *Lessons on Experimental Philosophy* (1748), a French work that went through several editions. The large pyramidal case was typical for storing compound microscopes.

152. Solar microscope. Used for large projections, the solar microscope was attached to a metal plate that could be mounted on a window shutter. A large mirror placed outside the window, but adjustable from the inside, reflected sunlight through the barrel of the microscope, which contained condensing lenses, a specimen slide, and an objective lens. In a darkened room, a considerably enlarged image of the specimen could thus be projected onto a screen or wall at some distance from the instrument. This diagram is taken from the *Philosophia Britannica* (1771) of Benjamin Martin, one of the outstanding instrument makers of eighteenth-century England and author of many books explaining scientific principles. Solar microscopes continued to be used until well into the nineteenth century. Charles Francis Adams, U.S. ambassador to England during the Civil War, recorded his introduction to a solar microscope at Harvard in 1824: "The magnifying power . . . was astonishing, exhibiting the finest fibres of a small portion of the finest of animals, that is to say, the most delicate. The lecture was a beautiful one and drew a much larger audience than usual, although the students could not avoid showing their boyish propensities."

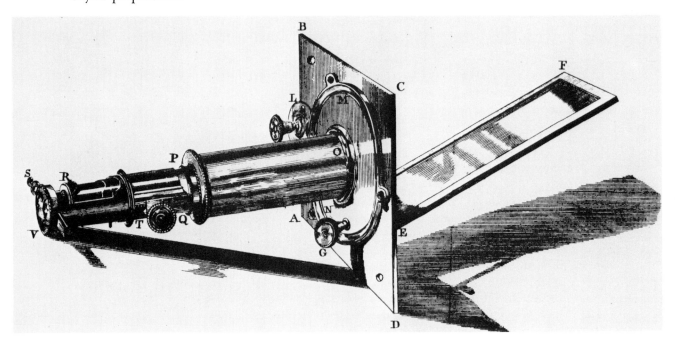

153. Lucernal microscope. Like the solar microscope, the lucernal microscope—invented by George Adams—was used to project an enlarged image of a slide or an opaque object. The light source was a lamp and a two-lens condenser system, as seen on the left in this drawing by a Mr. Cuthbertson sometime around 1790. The image appeared on a sheet of frosted glass at the end of a pyramidal box rather than on a screen or wall. The peep sight at the right ensures that the observer's eye is placed at the correct distance and in the proper position to view the enlarged image.

154. A royal microscope. In 1761 the English instrument maker George Adams constructed a silver microscope for King George III. Adams later described this instrument as follows: "It was made for His Majesty in the year 1761, and with it was then presented a manuscript of its use. . . . The [micrometer] screw has fifty-five threads to an inch, it carries an index pointing to the divisions on a circular plate fixed at right angles to the axis of the screw; its revolutions being counted on a scale of one inch divided into fifty parts." The microscope, intended for royal use, is built on a silver-cased Corinthian pillar, a little over a foot high; on each side there are two short pedestals, each with an ornamental vase or urn. The microscope barrel is supported by two silver figures. This instrument, which can be used as either a compound or a simple microscope, has been described as "the most artistically and elaborately decorated instrument ever constructed," exhibiting "the sacrifice of usefulness to ornament."

155. An eighteenth-century lodestone. In the seventeenth and eighteenth centuries, natural magnets, or lodestones, were studied at length to determine the nature of magnetism and the cause of the magnetic properties of the earth. As in this eighteenth-century Russian example, lodestones were often encased in elaborate brass or silver mountings, adorned with figures. At the bottom of the lodestone, two iron feet, or centers of concentrated or strong magnetic effects, protrude from the case. The "keeper," of which the main element is an iron bar with a hook, is held by magnetic force to the feet. Weights could then be suspended on the hook to demonstrate the lodestone's powers or to make quantitative measures of its strength. A load that exceeded the "force" of the lodestone would cause the keeper to fall off.

CHAPTER

14

The Experimental Physical Sciences

During the seventeenth and eighteenth centuries, many new areas of physical science were opened to experimentation, among them magnetism, pneumatics and hydrostatics, heat, sound, and electricity.

The beginning of the study of magnetism as an independent branch of physics was marked by the publication of William Gilbert's treatise *De magnete* . . . in 1600. Gilbert's preface expresses his point of view about the experimental study of nature: "This nature knowledge," he said, is "almost entirely new and unheard-of." He dedicated his work "To you alone, true philosophizers, honest men, who seek knowledge not from books alone but from things themselves."

Gilbert demonstrated the conditions under which iron can be magnetized and demagnetized, and he studied a number of aspects of geomagnetism. He came to the startling conclusion that the earth itself is a large spherical natural magnet, with a north and a south magnetic pole. Convinced that a spherical magnet would rotate, as he believed he had

found by experiment, Gilbert became a semi-Copernican—that is, he advocated the Copernican idea of the daily rotation of the earth, but was silent on the question of the earth's annual revolution around the sun, which was not germane to his experimental studies of magnetism.

Gilbert is also considered a founder of the science of electricity, since a whole chapter of his book was devoted to the electrical attraction of such objects as amber. He introduced the subject of electrical attraction primarily because he had found considerable confusion in the literature between electric and magnetic attractions, presumably by authors who had never actually seen both kinds of attraction.

One of the most spectacular subjects explored by experiment in the seventeenth century was that of air pressure and the properties of the vacuum. The pioneering work was done in Italy by Gasparo Berti, Evangelista Torricelli (a disciple of Galileo), and others who showed that the old principle that "Nature abhors a vacuum" is not sufficient to explain the action of pumps and other kindred phenomena.

Galileo reported that suction pumps are limited in their effect to a height of about 32 feet above water level. It would thus appear that Nature's alleged abhorrence was limited to 32 feet, which is obviously absurd. Torricelli introduced the concept that the earth is surrounded by a sea of air whose weight exerts pressure on it. He explained the action of pumps and their limitation to 32 feet of water in terms of air pressure. He reasoned that since mercury is fourteen times heavier than water, the air pressure that supports a 32-foot column of water would only support a 28-inch column of mercury (28 inches $\times$ 14 $=$ 32 feet). Experiment showed the correctness of this conclusion.

Blaise Pascal in 1648 devised a test. If the height of the mercury column in Torricelli's experiment is caused by (and corresponds to or measures) the pressure of the air, then the height of the column ought to decrease as one ascends a mountain; the reason is that the higher one goes, the less atmosphere or air there is above one to push down and cause air pressure. In Pascal's experiment, his brother-in-law ascended a mountain and measured the height of the mercury column at regular intervals. He found that the height of the mercury column did indeed drop just as Pascal had predicted. In order to make certain that the change in the height of the mercury column was caused only by the increase in the altitude of the observer and not by variations in air pressure during the day, Pascal set up a Torricellian mercury tube at the base of the mountain so that the height of the column could be observed from time to time during the day. This appears to have been the first "control" experiment on record.

Because the Torricellian tube measures the pressure or weight of the atmosphere, it became known as a barometer, from the Greek words *báros* ("weight") and *métron* ("measure"). Such instruments became common in the later seventeenth century and their use showed a correlation between variations in atmospheric pressure and changes in weather, as a result of which the barometer was called a "weather glass." It is this latter sense that we usually have in mind today when referring to a barometer. Experiments on air pressure led also to the invention of the air pump or vacuum pump.

In the study of heat, the dominant theory during most of the eighteenth century was that many thermal phenomena are caused by the motion of an invisible and imponderable fluid, which was eventually termed "caloric." Electrical phenomena were also explained by a similar imponderable fluid, and by the end of the century, most scientists believed that there are two imponderable electrical fluids. In the final years of the eighteenth century, the experiments of Benjamin Thompson, Count Rumford, were beginning to show the need for a wholly different kind of theory, based on the concept of heat as a mode of motion. We still continue to use the language of the old "fluid" theory of heat, however, in such expressions as heat transfer, heat capacity, heat flow, and heat conduction.

The eighteenth century was witness to many new concepts of heat, such as latent heat and specific heat. Experiments were made on the means of producing heat and of transferring heat, and measurements were made of the heat processes associated with the life process. The steam engine, harnessing the enormous power of heat, marked the beginning of a new era in human history, a technological revolution of which the end is still not in sight.

156. Gilbert's experiments on magnetism. Our modern science of magnetism dates from 1600, when William Gilbert published the founding treatise on the subject. One of Gilbert's main conclusions was that the earth is a great magnet with north and south magnetic poles. Gilbert's ideas were extremely important for Kepler, who assumed that the revolution of the planets in their orbits around the sun is the result of magnetic forces. In this illustration, Gilbert shows that if an iron bar is heated and hammered while placed in an appropriate north-south orientation, it will become magnetized with a pole at each end.

157. Berti's experiments with air pressure. In the 1640s Gasparo Berti made a series of experiments, apparently inspired by a discussion in Galileo's *Two New Sciences* (1638), which examined the reason why water cannot be raised more than 18 cubits (approximately 30 feet) by means of a lift pump. In Berti's experiments a long vertical tube was closed with a stopcock in the lower end, which was submerged in a tub of water. The tube was then filled with water and the top sealed tight. When the stopcock at the lower end was opened, Berti found that the level of the water in the tube dropped to about 30 feet. In the experiment shown here, Berti has enclosed a bell and clapper in the upper part of the sealed tube so that, after the water level has fallen, the bell and clapper would be in a "vacuum." When the clapper struck the bell a sound could be heard, perhaps transmitted through the metal attachment of the bell to the tube. Later experiments showed that sound is not transmitted through such a "vacuum."

158. The great Magdeburg experiment. In the 1660s, Otto von Guericke, the burgomaster of Magdeburg, in Germany, experimented with vacuums, using an air or vacuum pump he had invented. To demonstrate the tremendous force of air pressure, Guericke evacuated the air in two large hemispheres that fit together. When he attached a team of eight horses to a ring on each hemisphere, the combined force of the sixteen horses could not pull the hemispheres apart. Yet, when Guericke opened the stopcock and admitted ordinary air into the hemispheres, they could easily be separated.

159. The force of air pressure. The glass globe (left side of the balance) has been evacuated; the difference in weight between the evacuated globe and the same globe in its normal condition (i.e., full of air) gives the weight of the air the globe normally contains. In this experiment, a bottle full of air is connected to the stopcock at the bottom of the evacuated globe. When the stopcock is opened, air rushes into the globe, and the pressure of the circumambient air breaks the bottle. The globe can resist the pressure of the outer air, even when evacuated, because of its spherical shape, but the bottle, with four flat sides, is broken by the force of the outer air as soon as it loses some of its own air to the globe.

160. Boyle's experiments with air pressure. Robert Boyle, a gentleman-scientist of the seventeenth century, invented an improved air pump and conducted many chemical experiments. He employed Denis Papin and Robert Hooke as his assistants. Papin was later noted for the invention of the pressure cooker. Hooke was a great experimental scientist who, among other things, discovered the law of elastic force. In one experiment, Boyle fastened a large water pipe to the outside of a four-story brick house. Part of the pipe was made of a thin length of glass, so its contents could be observed. The bottom of the tube was submerged in a cistern of water, and the top was connected to a vacuum pump. When the pump was put into operation, the water began to rise in the pipe; an assistant, standing on the ground, added water to the cistern as needed. As in the experiments of Berti, the water would rise in the tube only to about 30 feet, despite the amount of pumping. The experiment demonstrated that the height to which the water could be lifted by a vacuum pump was limited by pressure from the surrounding air.

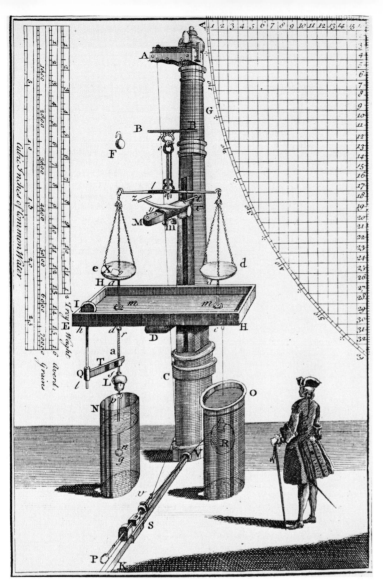

161. Hydrostatic balance. This elaborate eighteenth-century device was intended to determine the specific gravities of fluids and of solids. The general principle of the machine is simple—an application of Archimedes's laws of buoyancy to the weighing of substances in air and in water or some other fluid. According to these laws, a body immersed in a fluid (including air) is buoyed up by a force equal to the weight of fluid it disposes. It is not at all certain that the scale shown here is intended to represent accurately the size of this instrument.

162. The resistance of fluids. Although known today primarily for his work on the history of mathematics, Charles Bossut made important experiments in applied mechanics, the most notable of which were his studies of fluid resistance. He used an experimental basin and a ship model, as shown here, to test the resistance of water against the hull of a boat. Bossut was also famous for his rigorous textbooks that taught scientific principles to engineers.

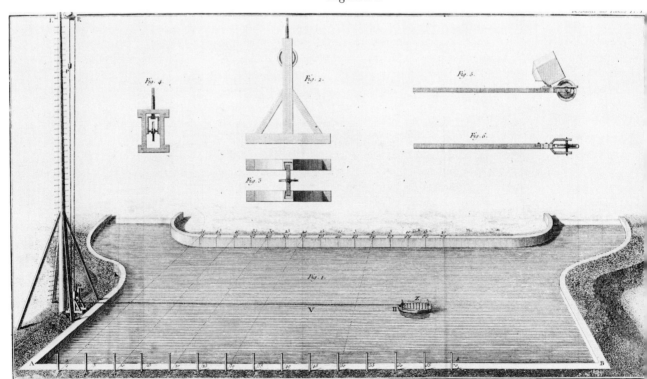

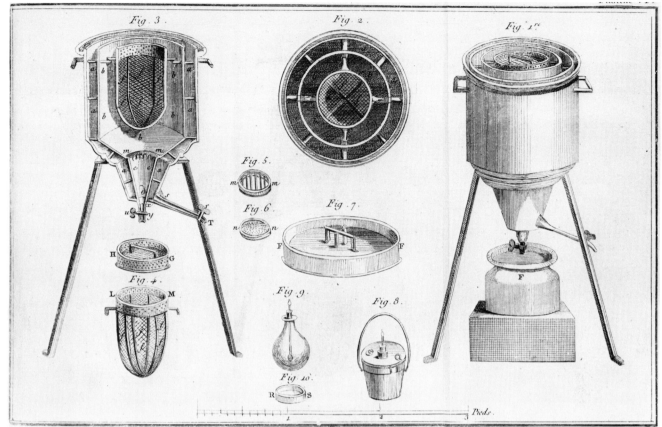

Paulze Lavoisier Sculp.

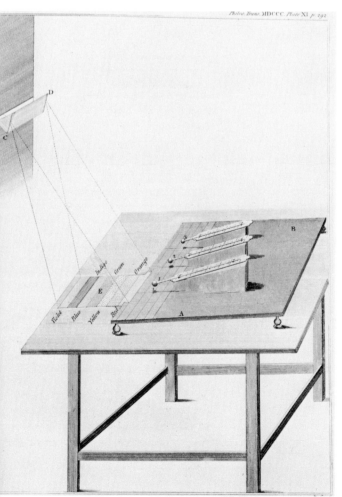

Philos. Trans. MDCCC. Plate XI p. 292

163. The ice calorimeter of Laplace and Lavoisier. This device has been described as the "earliest apparatus devised to measure the heat given off by a body, as distinct from measuring temperature." Essentially it measures heat in terms of the amount of ice melted by a hot body that cools off, by a chemical reaction that gives off heat (exothermic), or by a living animal giving off heat to a cooler surrounding environment. A guinea pig could be placed in the apparatus for several hours. In each case, the heat given off is measured in terms of the weight of ice that has been melted as a result. The heat could be accurately determined since it was known how much heat is required to melt a given quantity of ice. This illustration is based on one of the thirteen copper plates made by Mme. Lavoisier from her own sketches for Lavoisier's *Elementary Treatise on Chemistry* (1789). It is signed "Paulze Lavoisier sculpsit" ("engraved by [Mme. Marie Anne Pierrette] Paulze Lavoisier").

164. The nature of solar heat. In 1800 William Herschel published an *Investigation of the Powers of Prismatic Colours to Heat and Illuminate Objects*. With the apparatus shown here, he placed thermometers at different points on the visible spectrum in order to discover the comparative heating properties of different colored light emanating from the sun. Herschel also discovered that the spectrum extends into a region beyond the red, thus calling the attention of the scientific world to infrared radiation.

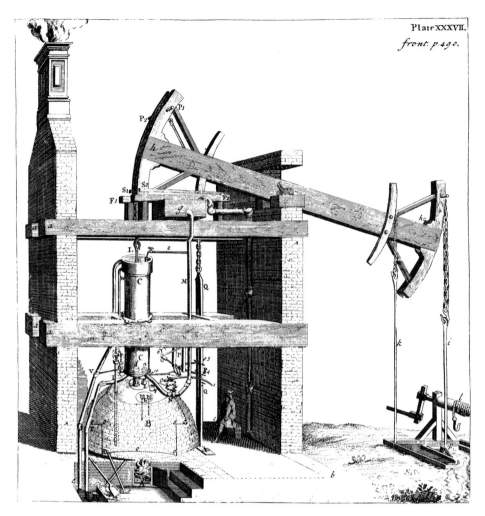

Plate XXXVII.

front: p.490.

165. Newcomen's "ENGINE for Raising Water [with a power made] by Fire."
The first practical piston steam engine was invented by Thomas Newcomen in
the early 1700s. The Newcomen engines were used primarily for pumping water
out of mines. This 1744 illustration shows a huge beam, pivoted at the center;
the pump rod is attached by a chain to the right end of the beam while another
chain at the left end links the beam to the piston of the engine. Steam is generated in the large hemispherical boiler, enters the vertical cylinder directly above
the boiler, and causes a piston inside the cylinder to be driven upward. The great
weight of the pump rod then tilts the beam to the position shown, the end of the
stroke. At this point a valve is closed so that no more steam is admitted into the
cylinder, and another valve is opened so that a stream of cold water is injected into
the cylinder, causing the steam to condense, thus producing a partial vacuum. The
pressure of the atmosphere then pushes the piston down again, rocking the beam
so as to pull up on the pump rod and produce a working stroke. The water had
to be drained out of the cylinder before steam was readmitted. Because this
engine worked on the action of the pressure of the atmosphere on the piston, it
was known as an atmospheric engine. Hundreds of such engines were built and
used in England during the eighteenth century.

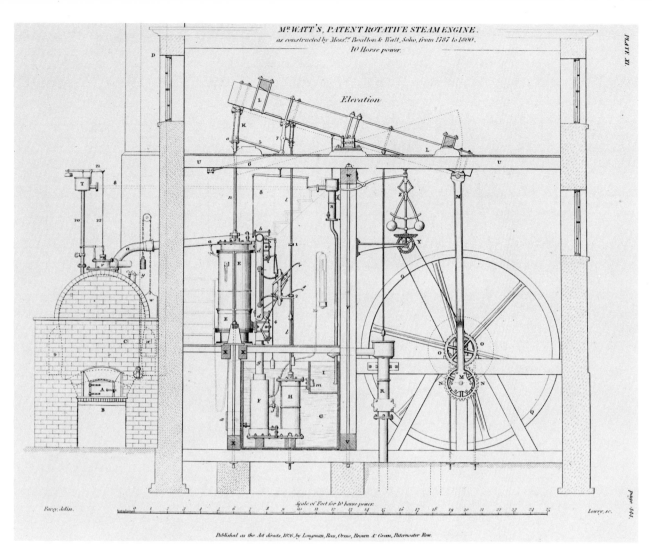

Published as the Act directs, 1826, by Longman, Rees, Orme, Brown & Green, Paternoster Row.

166. James Watt's steam engine. Watt's invention began with a radical transformation of the Newcomen engine. A series of heat experiments led him to introduce a "separate condenser," in which the steam could condense into water, since he found that to cool down the cylinder and piston and then reheat it by steam was very wasteful of heat energy. He made the engine "double acting"; that is, steam was admitted to drive the piston in both strokes. In the Newcomen engine, steam drove the piston in one of the two strokes, while the pressure of the atmosphere produced the returning or working stroke. Watt then found that even greater efficiency could be obtained by cutting off the steam before the end of the stroke, so that the final part of the stroke would be produced by the expansion of the steam already in the cylinder and under pressure. One of the most remarkable of Watt's inventions was the governor. This device consisted of a spinning or rotating system of two balls linked to rods. The speed of rotation of the governor was directly related to the speed of the steam engine. As the engine moved faster and faster, the two balls would move out further and further from the axis, closing down a valve to which they were linked, and thus cutting down the admission of steam and thereby slowing down the engine. This device made the steam engine a self-regulating machine, operating by feedback to maintain a constant speed—the first practical example of cybernetic principles in action. In the steam engine shown here, an ingenious gear system, known as sun-and-planets gearing, converts the reciprocating motion of the piston into rotary motion. Watt could not use the simpler device of the "crank" because of patent difficulties.

N. le Sueur Invenit   R. Brunet fecit.

167. Abbé Nollet's experiment on the electrification of human beings.
This experiment, of which a version was first performed by Stephen Gray
in England, is being demonstrated by Jean-Antoine Nollet to the ladies
of the French court in the mid-1740s. A young man, lying on a board
suspended from the ceiling by silk cords, is about to be electrified by
induction. The experimenter holds an electrified wand (in this case a
rubbed glass rod or tube) over the man's head, which causes a nonuni-
form distribution of charge in his body. When the woman raises her
finger near the boy's nose she draws a spark from it. After electrification,
the boy's hand attracts little bits of paper from a small table. Nollet was
one of the leading "electricians" of his day and published *Essai sur
l'électricité des corps* (1746), a treatise on his experiments from which
this illustration was taken.

15

Electricity– The New Science of the Eighteenth Century

Electricity is a relatively young branch of physical science. Unlike the subjects of statics, optics, motion, sound, and heat, it was virtually unknown to the Greeks. Nor was there enough known about electricity for it to constitute a legitimate branch of science in the Islamic period or during the European Middle Ages. Even in the age of Newton, the last half of the seventeenth century, knowledge concerning electrical phenomena was still so rudimentary that the distinction between conductors and nonconductors had not yet been made.

But a number of men had made observations about electricity. William Gilbert (see Chapter 14) studied the electrical attraction of various substances; Niccolò Cabeo observed electrical repulsion; Sir Thomas Browne coined the word "electricity"; Robert Boyle published the first work, a pamphlet, devoted wholly to the subject; and the members of the Italian Accademia del Cimento (Academy of Experiment), which flourished during the years 1657 to 1667, attempted to produce

electrification within a vacuum.

One of the most interesting seventeenth-century discoveries was the mutuality of attraction between an electrified (rubbed) body and the bodies it attracts. The discovery of this phenomenon has been traced back to Honoré Fabri. Otto von Guericke conducted some experiments by rubbing a sulphur ball that attracted and then repelled various light objects, but for him electricity was not a separate study so much as a part of his concern for questions of cosmology.

In the opening decades of the eighteenth century, major contributions to the knowledge of electricity were made by Francis Hawksbee, Stephen Gray, and Charles François de Cisternai Dufay. Hawksbee studied electrical effects within the evacuated space of an air pump and observed the electric glow; he also developed the concept of electric "effluvia." Gray studied the ways electricity (or electrification) may be communicated from one body to another, and he observed what came to be known as electrostatic induction. Dufay made the major discovery that there are two kinds of electrification, vitreous (as in glass rubbed with silk), and resinous (as in amber or wax rubbed with fur). In Benjamin Franklin's theory these states became associated with positive and negative charges. Dufay also found that bodies with the same electrification repel one another, while those with opposite electrification will attract one another (as will a nonelectrified body and a body with either electrification).

In Germany, meanwhile, various types of electrostatic generators were developed, and many spectacular forms of electrical experiments and demonstrations were conducted, including the ignition of warmed spirits and other combustible substances. In France, the major figure at this time was Abbé Jean-Antoine Nollet, who devised a theory of electrical action that was widely accepted.

As interest in electricity grew by leaps and bounds, contributions began to pour in from all over Europe. In England the leading "electrician" of the 1740s was William Watson. The most spectacular discovery of those years was the principle of the condenser, or capacitor, exemplified by the Leyden jar, a device first constructed by Ewald Georg von Kleist in 1745. Independently, this device was also produced in Leyden by Andreas Cunaeus and then by Jean Nicolas Sebastien Allamand in Petrus van Musschenbroek's laboratory. Musschenbroek reported the experiment of the Leyden jar to Nollet and it then became universally known.

The incredible aspect of the Leyden jar was that it could apparently store up and then suddenly release a prodigious quantity of electricity. By means of a Leyden jar, Louis-Guillaume Le Monnier was able to shock 140 of the king's courtiers, joined hand to hand in a continuous chain. Nollet repeated this experiment with 180 gendarmes and then with more than 200 Carthusian monks.

Then, in the late 1740s, a new theory was advanced in the New World by Franklin and his associates in Philadelphia. The basic feature of Franklin's theory of electricity is that every physical body, animate or inanimate, has a normal quantity of an electric fluid. Rubbing does not create electricity, but merely causes some of the original quantity of the electric fluid in one body to be transferred to another. Thus, one body will gain just as much electric fluid as the other loses. Franklin called the electrification of the body gaining excess electric fluid positive (or plus) and that of the body losing electric fluid negative (or minus). His theory is thus based firmly on the principle of conservation of charge—that all changes in electrification are the result of the appearance or disappearance of equal amounts of positive and negative charge.

Franklin's theory was very successful in explaining a large variety of phenomena, including the charging of a body by induction and the action of the Leyden jar, and it gathered a large following. His book on the subject was translated into French, German, and Italian. But the theory proved to have certain fundamental flaws; for example, it could not account for the repulsion between two negatively charged bodies. By the century's end— and despite the efforts of Franz Aepinus to amend it—his single-fluid theory was replaced by a theory of two electric fluids, one vitreous, or positive, the other resinous, or negative. The latter theory was espoused by Charles Augustin de Coulomb, who discovered and published the inverse-square law of electrical force and of magnetic force. Unknown to him, Henry Cavendish had already found the law of electrical force but had not published his results.

Franklin was particularly famous for his startling discovery that the lightning discharge is an electrical phenomenon. This knowledge, coupled with his discoveries about the nature of grounding and insulation, and the action of pointed and blunt conductors, led him to devise the lightning rod for the protection of houses, barns, public buildings, and even ships. At first, Franklin thought the lightning rod would merely draw off the "electrical fire" from passing clouds, but he later found that the lightning rod would also attract a stroke of lightning and conduct it safely into the ground or ocean. This was the first significant realization of Francis Bacon's promise that disinterested research into the workings of nature would lead to practical inventions of use to mankind.

By the end of the century, the subject of electricity had been extended from electrostatics (relating to the attraction, distribution, and repulsion of electric charges) to current electricity. This was made possible by Alessandro Volta's "crown of cups" and voltaic pile, early types of electric battery. With the battery, scientists had the means of producing

a sustained transfer of electricity in the form of a current, rather than just an instantaneous discharge like the spark or brush discharge in electrostatics. It was the electric current that made possible the electrical age that followed.

After publishing a paper on the battery in 1800, Volta went to Paris where he demonstrated his experiments and gave lectures at the Academy of Sciences. His lectures were attended by Napoleon, who was then First Consul. Napoleon, who was greatly interested in the sciences, proposed that Volta be given a gold medal. He personally authorized the Academy of Sciences to award a medal "for the best experiment made each year on the galvanic [electric] fluid" and a prize of 60,000 francs to the person who "by his experiments and discoveries makes a contribution to electricity and galvanism comparable to Franklin's and Volta's."

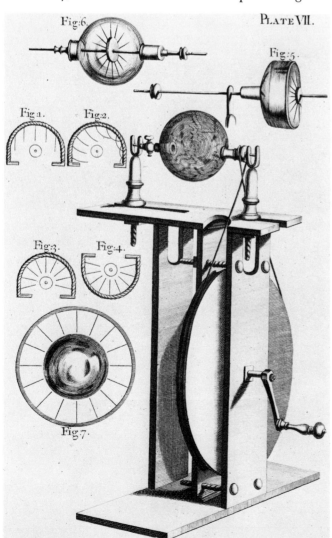

168. The electrical production of light. One of the phenomena that puzzled seventeenth-century scientists was the glow that occurred in a Torricellian vacuum (the space above the mercury in a tube containing some air at low pressure as well as mercury vapor) when the tube was shaken or jostled. Among the experiments performed in the early 1700s by Francis Hawksbee to investigate this phenomenon was one in which he caused a partially evacuated globe to rotate while holding his hand against it. The apparatus, shown here, was illustrated in his *Physico-Mechanical Experiments* (1714). Hawksbee reported that the light produced enabled him to distinguish the letters on the page of a printed book and to illuminate a wall ten feet away. In Figures 1 to 4, Hawksbee shows how a set of threads appeared when attached to a hoop placed around a glass cylinder that he spun and rubbed, as he had done with the globe; he believed that the configurations of the threads showed the direction of motion of the "effluvia," which he believed produces electrical effects. When he placed the threads inside the tube, they arranged themselves like the spokes in a wheel (Figure 7).

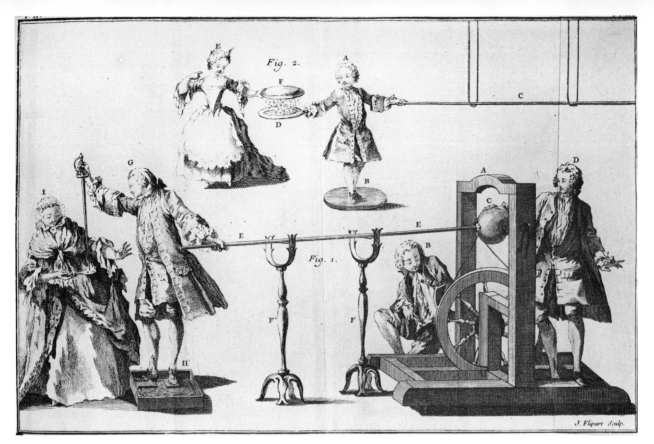

169. Electrical entertainment. In this illustration from a book by William Watson, a contemporary of Benjamin Franklin, electrification is being produced by a whirling globe in contact with an experimenter's hand (*right*). The electric charge passes along a metal tube, or prime conductor, that rests on two insulating stands. It then passes through another experimenter (*left*), who stands on an insulating wax cake, and through his sword to produce a spark that ignites some warmed alcohol in the spoon held by the lady next to him. The experiment originated in Germany. A second experiment (*top*) shows how bits of paper or straw are attracted and then repelled between two parallel plates, one held by a boy standing on an insulator and touching the prime conductor, the other held by a girl who stands on the floor (a ground) and whose plate receives a charge by induction.

170. Nollet's experiments on the electrification of plants and animals. In 1749 the Abbé Nollet published an account of his experiments on the possible effects of electrification on the growth of plants and on living animals. In this illustration, a chain touching the whirling globe of an electrostatic generator electrifies three connected shelves. In one experiment Nollet planted mustard seeds in two vessels, one of which he electrified for eight days; he reported that the electrified seeds "had all sprouted by the end of that time and had stalks fifteen or sixteen lines tall," but only "two or three of the nonelectrified plants had appeared above ground," and they had much shorter stalks. He also observed that an electrified cat lost slightly more weight than a nonelectrified cat, but he thought that this might have been caused by a "difference in temperament." Nollet also tested the possible medical or curative effects of electricity and concluded that there were none. Note that the water dripping from the can on the right jets horizontally on one side but is drawn down vertically on the other, in relation to the position relative to "the sphere of activity of the electrified globe."

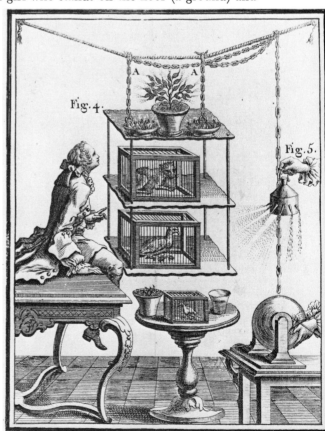

Fig.15.

Fig.14.

Expérience de Leyde

171. The experiment of the Leyden jar. The Leyden jar was the first condenser or capacitor; it consisted of a glass flask filled with water, into which a long wire was inserted and then electrified. Since the jar or flask was held in the experimenter's hand, the device actually consisted of two conductors—the water and wire together, and the experimenter's hand—separated by a nonconductor, the glass. This representation of the Leyden experiment is the earliest one known and is taken from Nollet's *Essai sur l'électricité des corps* (1746). The woman at the right is generating an electric charge by holding her hands on a whirling glass globe. The charge is then conducted through a long metal rod, or the prime conductor (which is suspended by two nonconducting silk cords), to a wire that extends into the water in the jar. When the first experimenters, Andreas Cunaeus, Jean Allamand, and then Petrus van Musschenbroek, attempted to withdraw the wire from the water, each experienced a severe shock. Musschenbroek wrote that he would not repeat the experiment "for all the kingdom of France," but he sent an account of it to Nollet for public dissemination. The upper part of the plate depicts Nollet's theory of electricity in which the same electrical ethereal matter moves in simultaneous effluent and affluent streams.

172. Nollet's experiment performed in Japan. One of the most spectacular electrical experiments of the mid-eighteenth century was set up by Nollet. It consisted of discharging a charged Leyden jar (or "battery" of Leyden jars) through some two hundred soldiers and a larger number of monks who formed a human chain. The shock of the discharge caused the participants to jump simultaneously. This depiction of a variation on Nollet's experiment was described in a book on medical electricity of 1813 and documents the introduction of Western science into Japan.

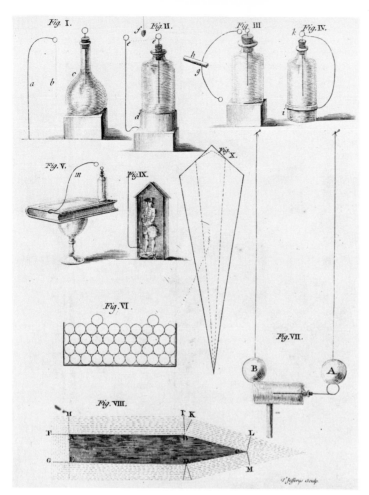

173. Franklin's fundamental experiments on electricity. One of Franklin's earliest and most impressive discoveries was the fact that a charged Leyden jar contains no more "electricity" than an uncharged one. He explained that when one of two conductors (say the wire and the water inside a glass jar) receives a positive charge by contact, the other conductor (either a metal coating around the jar or the experimenter's hand in contact with the jar) will gain a charge of the opposite sign by induction, if it is grounded. But the amount of the two charges, positive and negative, are the same. For example, in this plate from Franklin's *Experiments and Observations on Electricity* (1751), Figure II shows how a weight suspended on a silk nonconducting thread will oscillate between contacts with the inner and outer conductors of a charged Leyden jar until they are both wholly discharged. In Figure VII a charged Leyden jar with an insulating handle is moved so that its outer coating makes contact with ball *B* and the inner conductor's wire touches ball *A* and the jar is then removed; the two balls will have received equal and opposite charges. This will make the balls attract each other and touch, but since the amount of their charges of opposite sign are equal in magnitude, the balls will become fully discharged and separate again.

174. Franklin's sentry-box experiment. The wash drawing shown here was made under Franklin's direction for a manuscript copy of *Experiments and Observations on Electricity*. As Franklin envisaged his experiment, a box like the one seen here would be set on a high building. Through the box runs a stiff rod (or wire), pointed at the upper end and having the other end fixed into an insulating stand within the box. The tip of the rod would draw electricity from passing electrified clouds during a storm, and an experimenter, standing on the insulated stool inside the sentry box, would then be able to draw sparks from the rod, showing that the clouds had indeed been electrified and that electricity could be drawn safely along the rod into the ground. Soon after Franklin's *Experiments and Observations* had been translated into French, the experiment was performed successfully in France, but before Franklin heard news of it, he had already tested his idea by the lightning kite experiment and proved the electrical nature of lightning discharge.

21 To determine the Question, Whether the Clouds that contain Lightning are electrified or not, I would propose an Experiment to be try'd where it may be done conveniently. On the Top of some high Tower or Steeple, place a Kind of Sentry Box big enough to contain a Man and an electrical Stand. From the Middle of the Stand let an Iron Rod rise, and pass bending out of the Door, and then upright 20 or 30 feet, pointed very sharp at the End. If the Electrical Stand be kept clean and dry, a Man standing on it when such Clouds are passing low, might be electrified, and

175. Destruction of a church by lightning. On the right, there is portrayed a demonstration in which a Leyden jar is made to discharge in such a way that a spark is communicated to a knob on the steeple in the church model, thereby igniting some gunpowder inside and producing an explosion and a fire. On the left, an unprotected church, struck by lightning, illustrates what would happen in nature, as simulated in the demonstration model. If the little knob is removed, exposing a sharp metal point, as in a lightning rod, then there is a silent discharge, rather than a sudden surge and spark, and no explosion and fire. Thus it is shown that protection from lightning comes only from having a well-grounded lightning rod that ends in a point.

176. An explanation of atmospheric electricity. A diagram from a treatise by the Russian scientist M. V. Lomonosov illustrates the electrification of clouds and the discharge of lightning as a result of the friction from rising and falling currents of air.

177. A lightning experiment performed in Siberia. This engraving is from Jean-Baptiste Chappe d'Auteroche's account of his voyage to Siberia in 1761, undertaken to observe the transit of Venus. In the illustration, an insulated pointed rod has been erected in a field to attract an electric charge during a lightning storm; it is intended to measure the altitude of the lightning discharge. The apparatus permits the charge to be conducted along an insulated horizontal rod and collected in a Leyden jar. In a nearby shed the scientist is observing the phenomenon and making a measurement with his telescope, surrounded by a group of awed and apparently frightened Russian soldiers and peasants.

178. The largest electrical machine of the eighteenth century. This huge electrostatic plate generator was constructed by Martinus van Marum at the end of the eighteenth century, and was described by him in a book published in 1785. The machine still survives, and can be seen in the Teyler's Museum in Haarlem, the Netherlands. Some idea of its scale can be gained from the fact that the two rotating glass disks are more than 5 feet in diameter. Van Marum's experiments seemed to prove the validity of Franklin's "single fluid" theory of electricity. Alessandro Volta was convinced by van Marum's results and advocated a frontal attack on those scientists who held that electrical effects result from two "fluids."

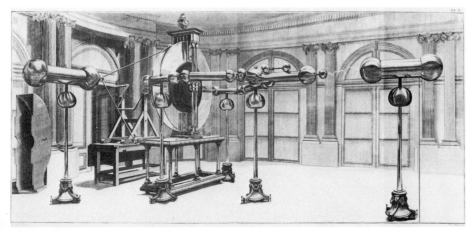

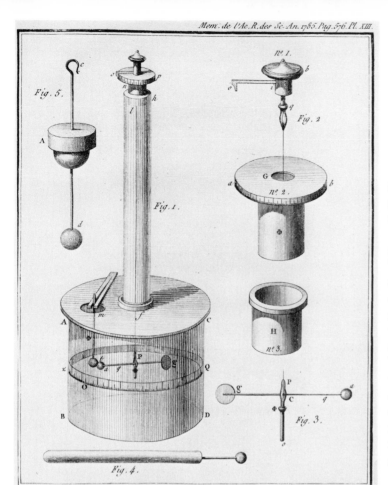

Fig. 5.

Fig. 1.

n.° 1.

Fig. 2.

n.° 2.

n.° 3.

Fig. 3.

Fig. 4.

179. The law of electrical force. In 1785 Charles Augustin de Coulomb published the results of his investigations concerning the force exerted between two small charged bodies. In the experiment shown here (Figure 1), he suspended a long stiff wire in a tube and from it hung a nonconducting bar with a small metal ball (*a*) at one end and a counterbalancing vane (*g*) at the other. The ball is charged and then brought into contact with a similar ball at the end of a nonconducting, upright rod clamped firmly in place, so that the two balls have an equal electric charge of the same sign. The resulting force of repulsion between the balls then pushes away the mobile ball, causing the wire to twist until the force of torsion equals the force of electric repulsion. The amount of force can be read on the scale at the top of the central column. Coulomb found that the law of electric force, like the law of magnetic force, took a form similar to Newton's law of universal gravitation, and that all three forces vary inversely as the square of the distance. (The other illustrations are details of Figure 1.)

180. Animal electricity. This wash drawing was made in preparation for an engraving that illustrates Luigi Galvani's *On Electrical Forces in Muscular Motion* (1791). It depicts a number of electrical experiments with partially dissected frogs' legs in which the crural nerve is exposed. Galvani discovered that if an experimenter holds a scalpel or piece of metal set in an insulating handle and touches the nerve with the metal, the leg of the frog will twitch whenever a spark is produced from a nearby electrostatic generator. He attributed this phenomenon to a kind of "animal electricity" in the tissue of the frog that was put into action by the electric spark. Later he found that a dissected leg will twitch if an "arc" of two different metals joins the nerve to the muscle.

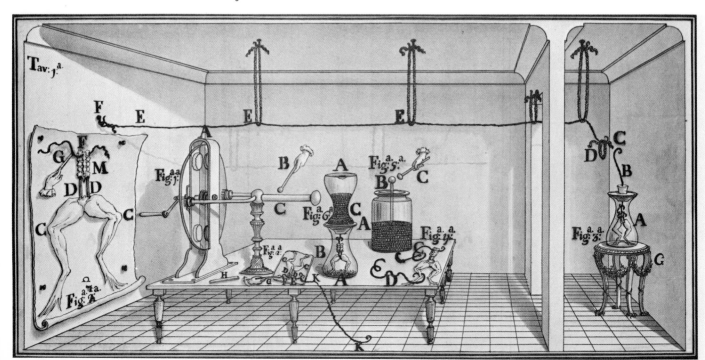

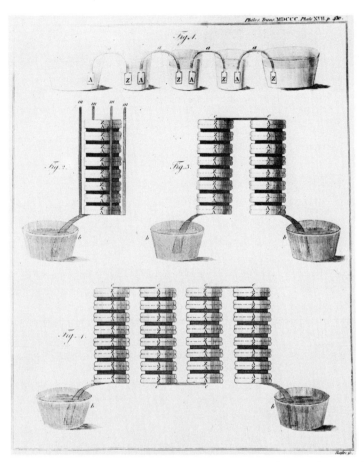

181. The discovery of the electric battery and electric current. Volta made a series of experiments to find out whether the electricity that produces the effects observed by Galvani is a special kind of animal electricity or the ordinary kind of electricity discussed by Franklin and others. In his attempts to answer this question, Volta showed that a steady electric current can be produced from two different metals, in this case zinc (Z) and copper (A), when they are partly immersed in a saline solution and then connected by means of a connecting rod or wire. A number of such wet cells could be duplicated and connected to form a "battery" in a configuration known as a "crown of cups" (Figure 1). Figures 2, 3, and 4 illustrate how sandwiches of zinc and copper could be separated by leather pads soaked in saline solution and stacked to form voltaic piles, or batteries of cells. In today's "dry cells" (batteries) one of the metals forms the actual cup or can itself, the other metal is replaced by a carbon rod running down the center of the cup, and the space between them is filled with an inert material moistened with a saline solution.

182. Lavoisier and his wife (ca. 1788). Jacques-Louis David's painting of Lavoisier and Marie-Anne Pierrette Paulze, his wife and collaborator, has been hailed as a masterpiece of portraiture, but the scientific apparatus in the picture betrays the artist's ignorance of physics. The mercury column is shown with a flat, rather than curved, surface. In 1773, Lavoisier wrote in his notebook that his program of research seemed to him bound "to produce a revolution in physics and in chemistry."

16

The Chemical Revolution

Chemistry emerged only gradually as an independent science. During the seventeenth century, it was cultivated in relation to alchemy, mineralogy, metallurgy, and pharmacy or medicine. Naturally, many scientists of the period were concerned with chemical topics, among them Robert Boyle, who is sometimes referred to as the "father of chemistry," but it was only during the last half of the eighteenth century that the subject came into its own and was given a form that we can easily recognize today. Antoine-Laurent Lavoisier was the chief architect of the chemical revolution, and he described his own work as revolutionary.

One feature that distinguished the chemical science of the eighteenth century from previous ages was the development of methods of studying gases. A valuable tool for research in gases was the pneumatic trough, invented by Stephen Hales and developed to a high level by Joseph Priestley and Lavoisier. Priestley, a nonconformist British clergyman and a friend of Benjamin Franklin, was an ardent

supporter of the American and French revolutions who emigrated to America in 1794.

One of Priestley's major discoveries was oxygen, which was also discovered two years earlier by Carl Wilhelm Scheele in Sweden. Priestley found oxygen to be eminently suitable for breathing and capable of supporting combustion. Since the theory then in vogue held that combustion, or burning, was a process in which a principle of heat, or "phlogiston," is given off, Priestley called the gas he had discovered "dephlogisticated air." He found that mice could breathe this "dephlogisticated air" and that candles would burn in it with particular brilliance. Describing his work in *Experiments and Observations on Different Kinds of Air* (1774), he said:

> My reader will not wonder that, after having ascertained the superior goodness of dephlogisticated air by mice living in it, and the other tests above mentioned, I should have the curiosity to taste it myself. I have gratified that curiosity by breathing it. . . . The feeling of it to my lungs is not

sensibly different of that from common air, but I fancied that my breath felt peculiarly light and easy for some time afterwards. Who can tell but that, in time, this pure air may become a fashionable article in luxury? Hitherto, only two mice and myself have had the privilege of breathing it.

It was Lavoisier who realized the full import of Priestley's discovery. In Lavoisier's new system of chemistry, Priestley's "dephlogisticated air" was not an altered form of the "element air," but a gaseous chemical element, oxygen. Lavoisier called "oxide" the compound formed from the chemical union of a metal and oxygen. He showed that when an oxide is reduced by heating it with charcoal to give a metal, the oxygen in the calx or

oxide combines with the carbon to form a gas, leaving the metal free. Lavoisier also showed that oxygen is one of the components of our air, which is a mixture of gases. When a metal is heated in air, a calx is formed by a chemical combination of the metal and the oxygen in the air.

Lavoisier's explanation of combustion, calcination, and other processes was eventually combined with studies of human respiration. A guiding principle of his chemical work was that the total weight of all the reacting substances must always equal the total weight of all the products. This is sometimes known as the law of conservation of matter; it is equivalent to saying that matter (mass or weight) is neither created nor destroyed in chemical reactions.

183. **Distillation furnace of the late fifteenth century.** A compound furnace with a vent at each corner, such as the one in this woodcut, ensures that heat will be drawn equally from the central chimney into the four separate stills at the corners. At the left, an apprentice is collecting the distillate from a protruding alembic, or distillation apparatus, which is similar to those used in alchemical experiments.

Ro: Vaughan sculp:

184. The chemical balance. Thomas Norton's *Ordinal of Alchimy* was begun in 1477 and written as a rhymed poem. It was not published in English until 1652, when it appeared in Elias Ashmole's collection entitled *Theatrum chemicum Britannicum*; but a Latin edition, printed in Germany in 1618, was widely disseminated. Newton is known to have carefully studied the work. The most interesting aspect of this illustration from the book is the fact that the furnaces, retorts, and other apparatus are actually chemical. The figure at the desk is presumably Norton himself, and this may be the first representation of a balance enclosed in a case.

185. The calcination of antimony. This plate, taken from an English translation of Nicaise Le Febvre's *Compleat Body of Chymistry* (1664), shows a "Chymical Artist" calcining antimony, that is, using heat to convert powdered antimony into its oxide. The heat is provided by the sun, whose rays are focused on the antimony by means of a large burning glass. The light produced by the reaction is so bright, the experimenter must protect his eyes with dark glasses.

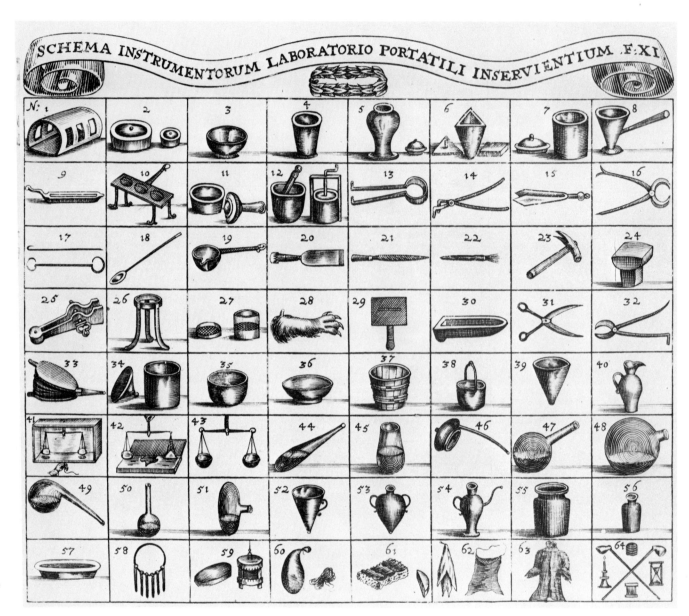

186. Becher's "portable laboratory." In 1689 the chemist Johann Joachim Becher cataloged in words and pictures the equipment required by a chemist for his work. Included were mortars and pestles (11 and 12), various tongs (13 through 16), a bellows (33), and three different types of balance (41 through 43), one of which is obviously a sensitive analytical balance in a glass case.

187. The discovery of phosphorus. Joseph Wright of Derby, England, artistically recreated the chemistry of a bygone time in this 1771 painting. An aged chemist kneels in wonder before the brilliant light being emitted by phosphorus, a new and radiant element. Its light is contrasted with the light of the candle on the table at the left and with the light of the moon shining through the window; both are pale by comparison. Phosphorus was actually discovered by a German merchant, Hennig Brand, who took up the pursuit of alchemy. In 1669, after distilling enormous quantities of urine, he obtained powdered phosphorus, which glowed in the dark with a continuous white light. He was the first person in recorded history to discover a chemical element.

188. A seventeenth-century view of a chemist's laboratory. This scene, one of a series of engravings based on a painting by David Teniers the Younger, shows an aged chemist at work at his furnace. He has three young assistants, some of whom may be apprentices. Such prints were popular all over Europe at the time, indicating the widespread interest in chemistry.

LE CHIMISTE

Dedié à Monsieur Jacques Jean Comte de Wassenaer   Seigneur d'Obdam, Chevalier du Saint Empire Romain.

189. Chemical laboratory of the Paris Academy of Sciences (1676). This engraving by Sebastien Le Clerc was published in *Memoirs on the History of Plants*. It shows scientists crowded around a table conducting chemical experiments and, outside the building, gardeners cultivating plants. The implication of the picture, borne out in the text of the book, is that the study of chemical processes will provide a clue to nature's operations in the animate world.

190. Practical chemistry in the mid-eighteenth century. In 1747 the *Universal Magazine* published this view of "Practical Chymistry." The chemist is shown as an aged scholar, giving instructions to a workman. The most notable aspect of the room is the collection of furnaces and stills. The various stuffed animals hanging from the ceiling include a crocodile, a beast traditionally associated with alchemy.

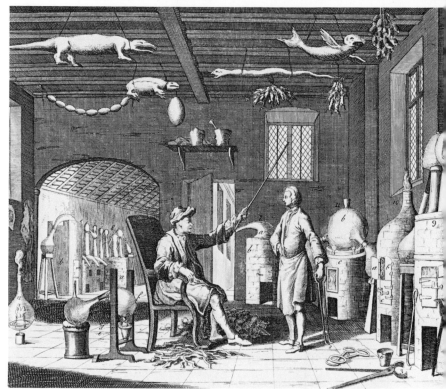

191. French laboratory of the eighteenth century. This view of the laboratory of the Capuchin monks in the Louvre includes a characteristic collection of glassware, stills, and retorts. The Latin inscription on the chimney reads *"nec pluribus Impar"* ("I suffice for more worlds than one"). With it appears the emblem of the sun, for Louis XIV, the Sun King. The inscription on the floor is from the Apocrypha, Book of Esdras, and reads in translation, "Measure for me a blast of wind, weigh for me a pound of fire, and call back a day that has passed." The quote from Psalm XIX along the bottom of the picture declares, "Day unto day uttereth speech, and night unto night sheweth knowledge."

192. Priestley's apparatus for investigating the chemical composition of the atmosphere. Dominating this illustration from Priestley's *Experiments and Observations on Different Kinds of Air* (1774) is a large earthenware pneumatic trough, 8 inches deep. At the far end are flat stones that hold inverted jars for collecting gases. The inverted glass with a stem is a tall beer glass (d), containing air sufficient to support the breathing of a mouse for about a half hour, and "something" on which the mouse, which can be seen inside the glass, "may conveniently sit, out of reach of the water." Priestley described how he introduced the mouse into a glass full of oxygen (which he called "dephlogisticated air") by passing it rapidly through the water. In the foreground a small jar with a perforated top stands on a perforated plate. Mice were kept alive in it for experiments to determine whether the new "air" could support life. Priestley said that to avoid chilling the mice, he kept the jar and stand on a shelf over his kitchen fireplace. A bent rod with a small candle (12) leans on the stand holding the jar with the mice. This was to permit the rapid insertion of a burning candle, with the flame upward, into a vessel filled with the new "air."

193. Priestley's laboratory. In his *Experiments and Observations*, Priestley also showed a portion of the room that he had converted into a laboratory. The apparatus in the fireplace was used to expel gas from certain solids, which were put into a gun barrel placed within the fire. The open end of the gun barrel was luted to the stem of a tobacco pipe, which led into a trough of mercury suspended from the mantelpiece. In this trough there is an inverted cylinder, full of mercury, in which the gas is collected. At the end of the table nearest the fireplace, gas is being generated in a test tube heated by a candle; the gas then passes through a small glass trap to remove moisture and is then collected in an inverted glass tube full of mercury set in a basin of mercury. At the other end of the table a gas generator is attached by means of a flexible leather tube to a container filled with a fluid that will be impregnated with gas; using a flexible connecting tube made it possible for Priestley to shake the gas generator from time to time. On the round table in the left foreground, a quantity of gas is trapped in a large U-shaped tube, the open ends of which are standing in small vessels of mercury. This device, known as a mercury siphon, contains an iron wire in each leg. Priestley used this apparatus to experiment on the effects of electricity on gases; he could produce a spark in the gas or gases at the top of the U-tube or siphon, which would cause a reaction.

194. Lavoisier's table of the elements. Lavoisier's reform of chemistry is most clearly seen in his revision of chemical nomenclature, seen here in an English translation. He set forth simple operational terms for the definition of chemical elements, or "simple substances," and then named compounds in terms of the relative proportion of their constituent chemical elements. He eliminated such names as "dephlogisticated air," "phlogisticated gas" or "phlogisticated air," and "inflammable gas" or "inflammable air" and replaced them with the chemical names: oxygen, azote, and hydrogen. The name "oxygen" was coined from the Greek work *oxys* meaning *acid*, as a result of Lavoisier's mistaken idea that the element is a constituent of all acids. Azote came from the Greek *a* (for *not*) and *zoe* (meaning *life*) and designated nitrogen, because the gas does not support life. Previously these gases had been thought of as varieties of "air," but, in Lavoisier's conceptual scheme, they became chemical elements just like the metals. At the head of Lavoisier's list are five "simple substances belonging to all the kingdoms of nature." The first of these is light, followed by caloric, the fluid or principle of heat.

TABLE OF SIMPLE SUBSTANCES.

Simple fubftances belonging to all the kingdoms of nature, which may be confidered as the elements of bodies.

New Names.			Correfpondent old Names.
Light	-	-	Light.
Caloric	-	-	Heat. / Principle or element of heat. / Fire. Igneous fluid. / Matter of fire and of heat.
Oxygen	-	-	Dephlogifticated air. / Empyreal air. / Vital air, or / Bafe of vital air.
Azote	-	-	Phlogifticated air or gas. / Mephitis, or its bafe.
Hydrogen	-	-	Inflammable air or gas, / or the bafe of inflammable air.

Oxydable and Acidifiable fimple Subftances not Metallic.

New Names.			Correfpondent old names.
Sulphur	-	-	The fame names.
Phofphorus	-	-	
Charcoal	-	-	
Muriatic radical	-		Still unknown.
Fluoric radical	-	-	
Boracic radical	-	-	

Oxydable and Acidifiable fimple Metallic Bodies.

New Names.			Correfpondent Old Names.
Antimony	-		Antimony.
Arfenic	-	-	Arfenic.
Bifmuth	-	-	Bifmuth.
Cobalt	-	-	Cobalt.
Copper	-	-	Copper.
Gold	-		Gold.
Iron	-	-	Iron.
Lead	-	-	Lead.
Manganefe	-	-	Manganefe.
Mercury	-	-	Mercury.
Molybdena	-	-	Molybdena.
Nickel	-	-	Nickel.
Platina	-	-	Platina.
Silver	-		Silver.
Tin	-		Tin.
Tungftein	-	-	Tungftein.
Zinc	-	-	Zinc.

Regulus of

Salifiable

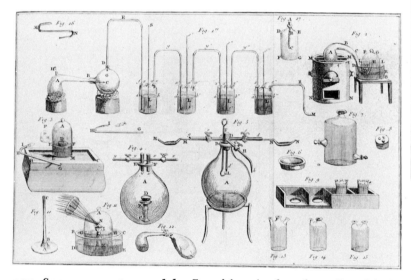

195. Some apparatus used by Lavoisier. A plate from Lavoisier's *Elementary Treatise on Chemistry* (1789) shows the varieties of glassware he used. This illustration, like others in the book, was drawn by Mme. Lavoisier. At left center (Figure 11) can be seen Lavoisier's apparatus for studying the combustion of phosphorus in a limited quantity of air. The phosphorus is placed in a small vessel on a pedestal inside a bell jar standing in a basin of water. The heat for combustion comes from the sun's rays, brought to a focus by means of a large lens that belongs to the Paris Academy of Sciences. Lavoisier performed this experiment over water and also over mercury, and discovered that in the course of combustion, the phosphorus used up about one-fifth of the volume of the enclosed air. At the upper right-hand side of the plate, in Figure 2, is Lavoisier's apparatus for studying the "calx" of mercury. Lavoisier's experiments showed that oxygen is the active "principle" (or chemical element) that causes an increase in weight during calcination.

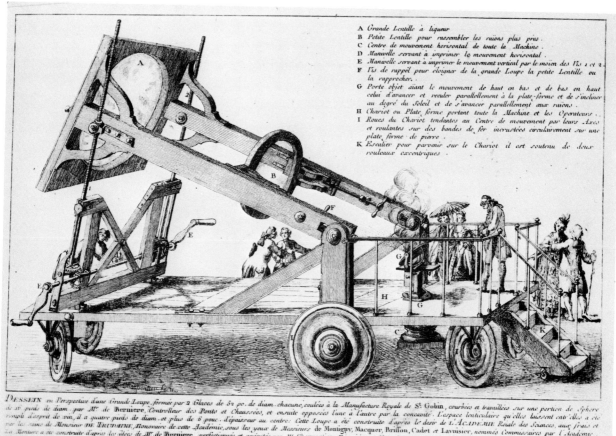

196. The great lens of the Paris Academy of Sciences (1782). This lens was constructed at the royal factory of St. Gobain under the direction of a committee of academy scientists consisting of de Montigny, Pierre Joseph Macquer, Mathurin-Jacques Brisson, Louis-Claude Cadet, and Lavoisier. Its purpose was to produce intense localized heat for performing chemical experiments, during which the experimenter had to protect his eyes with dark glasses. Lavoisier had previously used one of the academy's large lenses to study the combustion of a diamond. The heat of this great lens was not sufficient to melt platinum, but Lavoisier was able to produce fusion of platinum by heating it with burning charcoal in a stream of pure oxygen. Benjamin Franklin, who was in Paris as a diplomat at the time, called the fire produced by pure oxygen "much more powerful than the strongest burning mirror," and said it was "the strongest fire we yet know."

Part Five

THE
EARTH
SCIENCES

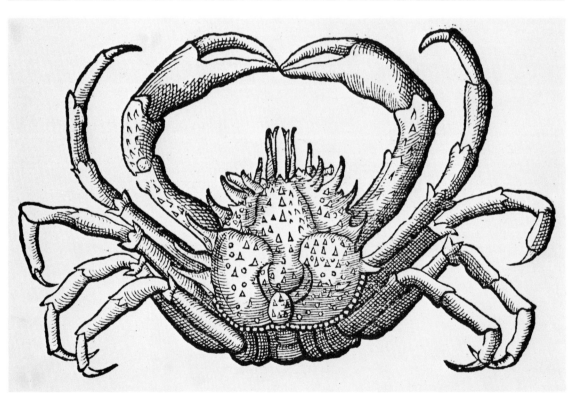

Pagurus la-
pideus, parte
supina expres
sus.
Ein steininer
Meerkrebß / o=
der Täschen=
krebß.

197. The beginning of scientific paleontology. In 1558 Konrad Gesner, the Swiss naturalist, published the first treatise in which illustrations are used as an integral part of the discussion of fossils. Martin Rudwick has pointed out that the importance of Gesner's innovations "can hardly be exaggerated." For the first time it was possible for a student of fossils to be "certain that he was applying a name in the same sense as his predecessors." Indeed, Gesner's own aim was to use illustrations "so that students may more easily recognize objects that cannot be very clearly described in words." This illustration of a fossil crab (*top*), together with a caption in German and in Latin, comes from Gesner's book on fossils (1565). The companion illustration of a living crab, Pagurus, which it strongly resembles, comes from Gesner's *History of Animals* (1558).

17

Studying the Earth

The sixteenth and seventeenth centuries were characterized by extensive exploration and geographical mapping. With increased world trade, and the establishment of colonies by European nations, new kinds of maps were required for navigators. Of these, the Mercator projection was of particular significance, since it enabled seamen to plot a course at constant angles to the meridian by drawing a straight line with the edge of a ruler.

During these centuries, there was much speculation about how the earth might have assumed its present form, on the supposition that the primitive earth was different. In part, these discussions arose from the discovery that the crust of the earth is not uniform, but is composed of layers or strata. One of the chief problems arose from the fact that the strata themselves are not uniform, but are tilted, fold over one another, and often project vertically with abrupt endings. In 1644 Descartes published an influential explanation of the formation of the earth as it now exists, based on a theory that assumed that the earth was

originally a hot "star" that eventually cooled off. He hypothesized that the crust that formed in the cooling process would have collapsed, falling into the liquid below and producing the irregularities of mountains, continents, and oceans. This theory was developed in an ingenious fashion by a Danish scientist, Niels Stensen, better known as Nicolaus Steno, who supposed that in the valleys formed when the horizontal strata collapsed, an inundation produced new sedimentary layers, which in turn might collapse. Both of these theories, which were closely related, assumed that the development of the earth could be explained by mechanical forces.

Well into the nineteenth century, most theories of the history of the earth were founded on the unassailable postulate that it had been created during the "six days" of Genesis, although there was some discussion as to whether these "days" might have been longer periods of time. Most writers also assumed the reality of the Flood. Many geologists supposed that the presence of fossils—

particularly marine fossils, widely distributed on the earth's surface and found even on mountains—was evidence that the surface had once been covered with water. The fossils were considered remains of aquatic animals and plants of a bygone day. Others, however, did not believe in the organic origin of fossils, but claimed that they were geologic formations that only had the external form or appearance of animals and plants.

By the eighteenth century, the issue of the animate versus the geological origin of fossils was decided in favor of the former. By the century's end, it had become clear that fossils represent species that are similar to, but not exactly the same as, species found among living plants and animals today. Eventually, there arose a theory of successive "revolutions" or "catastrophes" in which forms of life disappeared, leaving their fossil remains; afterwards, new forms of life would be created. Since the appearance of fossils in different strata of the earth is related to the time sequences in which these layers were formed, the dating of fossils became of great importance.

Georges Cuvier, one of the greatest naturalists who ever lived, and the true founder of the science of comparative anatomy, used his extensive knowledge of anatomy to clarify many of the fundamental issues concerning the age and history of the earth and of animal populations. He assembled fossil bones from America to form the skeleton of a mastodon, showing that a particular collection of bones all came from a single animal, rather than from parts of many different animals. He made anatomical comparisons of elephants from Africa and India, showing in significant detail how these differed from each other and how both differed from the Siberian mammoth. His studies on the size and bone structure of birds found in Egyptian tombs showed that there had been no essential changes in animal life during the few thousand years that separate us from Pharaonic times. Accordingly, he argued, the time required for the kinds of changes shown by the fossil record would have to be so long as to make the age of the earth far greater than had been commonly believed.

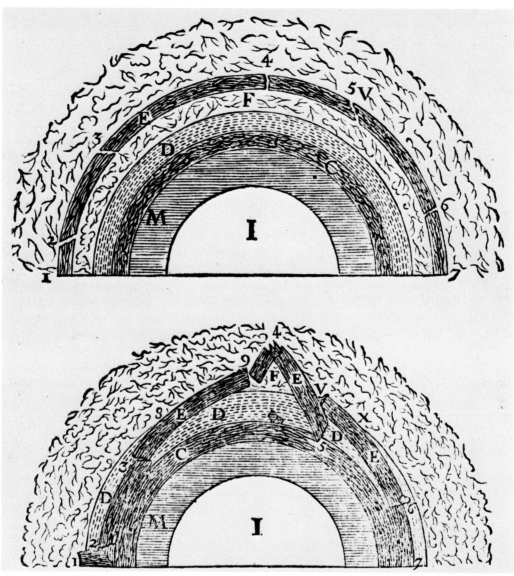

198. Descartes's theory of the formation of the earth and of mountain building.
In his *Principia philosophiae* (1644), Descartes stated that he would not deny
that the earth was created as we know it today. On the other hand, there is nothing
to prevent man's rational faculties from exploring hypotheses about how the earth
might have assumed its present form through a series of gradual natural processes.
He held that the earth, like the sun and stars, could have been formed of a glowing mass, which cooled and became compact. Descartes hypothesized three regions
of the earth: a central part containing the "first element" in rapid motion; a solid
and opaque region surrounding it; and a third region in which there are many
kinds of particles that are acted on by weight, light, and heat. With the passage of
time, the outer region is altered by evaporation and condensation, by the phenomenon of terrestrial weight, and by the heat and light of the sun. An atmosphere
of air is formed, the vapors of the air condense to form oceans, lakes, and rivers,
and then the phenomenon of tides begins. Fire, acting through gaps in the structure of the outer terrestrial matter, causes earthquakes and volcanoes. The harden-
ing and contractions of matter lead to a crustal motion that produces mountains,
as shown in the diagrams. Descartes's theory of the earth was significant because
it gave a mechanical explanation of the formation of the irregular surface of the
earth and showed why there are tilted strata that are neither horizontal nor curved
slightly so as to follow the shape of the earth. Descartes boasted that he had
described the development of the earth "as if it were only a machine in which
there is nothing whatever to be considered but the shape and movement of
its parts."

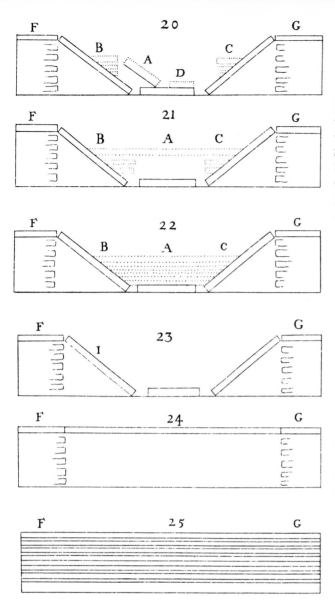

199. The first geological sections. Nicolaus Steno, a brilliant Danish physician and natural scientist, studied in Holland and then moved to Italy, where he carried out research in geology and became a member of the Accademia del Cimento in Florence. His *Prodromus* (1669) sets forth a theory on how the strata that compose the earth's crust may have given rise to mountains. His theories were illustrated in a series of six diagrams, based on his observations in Tuscany. The bottom diagram shows the strata unbroken and parallel to the horizon. The next one up illustrates how cavities are formed as fire and water eat away interior regions while the upper strata remain unbroken. Finally *(top)*, new hills and the breaking and collapsing of the upper strata, thus forming mountains and valleys. In the next diagram, the new strata are seen to arise from deposits from the seas in the valleys. Then, the lower strata of these new beds are destroyed by the action of fire and water, while the new upper strata remain unbroken. Finally, at top, new hills and valleys result when the new upper sandy strata are broken down in turn. The solid lines indicate what Steno called "rocky strata," while the dotted lines represent the "sandy strata" of the earth. These names came from the predominant element in each of the strata. Steno noted that the diagrams had been "geometricized" in order to simplify and illustrate his theory of the stages in the history of the earth.

200. A seventeenth-century theory of volcanoes. In a book entitled *Mundus subterraneus* (1665), the Jesuit scientist Athanasius Kircher presented his theory of the earth's structure. At the center he pictured a great fire, which sometimes reaches the surface via many fissures or pathways, resulting in volcanoes. Kircher also believed that there are caverns filled with water inside the earth. This water is heated by the central fire and rises to the surface in the form of hot springs. Like Descartes, Kircher sought to explain terrestrial phenomena in terms of large-scale natural forces.

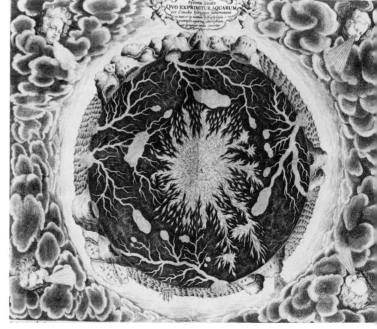

201. Mesozoic fossils. Martin Lister's *Historia ani-malium Angliae* (1678) was of great importance to the history of geology and paleontology, because of its splendid illustrations of fossils and recent or contemporary marine animals. Lister was interested in the general aspects of natural history, particularly conchology, and he was drawn to "shell stones," or fossils, by the controversy over whether fossils were really of animal orgin. Lister was so aware of the differences in detail between existing shells and fossil shells that he could not believe there were direct links between them. He made a strong argument for the geological rather than the animal origin of fossil shells based on his correlations of their distribution and certain types of rocks. In 1684 he made a suggestion to the Royal Society that geological maps be compiled.

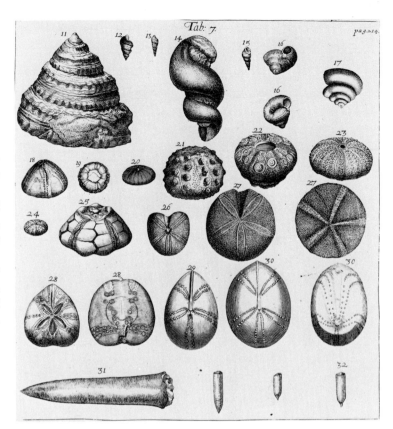

202. The Flood and fossils. The Swiss naturalist Johann Jakob Scheuchzer is considered to be the founder of paleobotany and of modern European paleontology. He made an extensive study of fossils and argued that they had an organic origin and were not geological formations. His *Complaints and Claims of the Fishes* (1709) had the fossil fish speak up in order to assert their organic origin and to confute those who have said that they were never alive. The fish declare that they were witnesses to the Deluge; thus, their very existence gives evidence of the reality of that event. It is the Flood that explains why their remains are to be found in high places where today there is no water. A similar theme occurs in Scheuchzer's *Herbarium of the Deluge* (1709), which contains many illustrations of fossil plants, together with accurate descriptions. In 1716 Scheuchzer published a catalog of his collection of fossils. In this illustration from the work, Scheuchzer himself is pointing to some of the different types of fossils he had found in the Alps. One of his students is digging with a pick, seeking additional specimens.

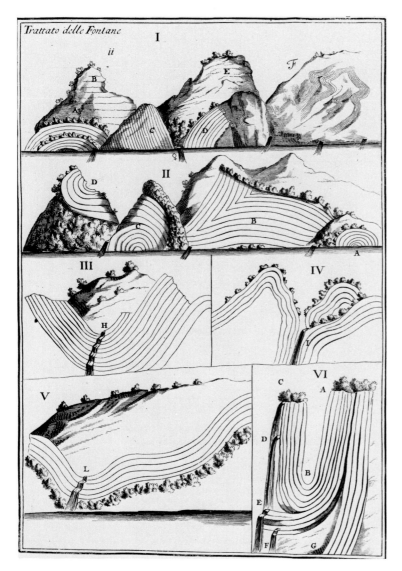

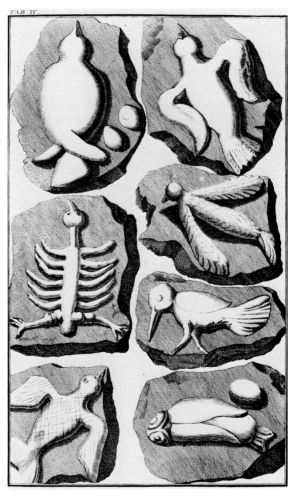

203. Geological sections according to Vallisnieri. Antonio Vallisnieri was interested in the origin of springs and of fossils. These drawings of geological sections, taken from his book on the origins of springs, published in Venice in 1715, were based upon his studies of the folded strata of the Alps and on drawings made and given to him by Scheuchzer. Vallisnieri was also noted for his investigations in natural history and zoology, and he studied the reproductive systems of man and animals, particularly those of the parasitic insects of plants. Although he denied the function of spermatozoa in fertilization, he did make an important confirmation of the general law that all parasitic insects of plants are derived from eggs.

204. The great fossil hoax. Johann Barthel Adam Behringer, a professor of medicine and dean of the faculty at Würzburg, collected fossil shells for his cabinet of natural curiosities and for his lectures on natural history. Two of his colleagues, a professor of geography and mathematics at the university and the university's librarian, together with three young boys, molded and baked "stones," so that they would exhibit a variety of unusual shapes, and then hid them about the region near Würzburg where Behringer searched for fossils. In due time he found the fakes, and—despite the efforts of the alarmed hoaxers to dissuade him—he published a report of his findings in 1726, entitled *Lithographia Wirceburgensis*, or *Graven Stones of Würzburg*. Plate 4 of his book, reproduced here, shows some of the fake animals on the stones. Other plates show stones with Hebrew letters, figures of comets and of the sun and moon, strange marine animals, bees in their hives sucking the honey from fossil plants, and birds in flight or sitting in their nests. According to Martin Rudwick, Behringer appears to have concluded that since the fossils he illustrated were only "imitations" of organisms, they provided evidence "for the inorganic origin of fossils."

Le Mont Blanc vu du pied de l'Aiguille Marbrée au N. E. de la Cabane .
A. Cime du M. Blanc. B, C. Aiguilles attenantes au M. Blanc du Côté de Cormayeur, D. monticule au pied du quel étoient les tentes,

205. Studies of the Alps. In a set of four volumes that appeared from 1779 to 1796 Horace-Bénédict de Saussure published an account of his researches on the geology of the Alps. In the illustration, the geologist, accompanied by a guide or assistant, is dwarfed by the immensity of the snow and ice around Mont Blanc; its summit is marked *A*. De Saussure's work has been described as "the first geological reconnaissance of the whole area of the Swiss Alps." In the fourth and final volume, from which this illustration is taken, de Saussure provided evidence that the formations he observed could not have been the result of mere crystallization from the waters of primitive oceans, but must have originated in the action of some explosive upward force or in a folding over of original horizontal strata. He could see no cause for an explosion, such as subterranean fires, and supposed that the various strata were folded over upon one another when the rocks were still in a plastic condition. "I can offer no general theory to account for the origin of these mountains," he wrote, "for which I await the results of further observations."

206. The Neptunian theory. This drawing by Abraham Werner appears in a 1786 essay on the classification and description of various rocks. Werner was the undoubted doyen of geologists during the 1770s and 1780s, and he pioneered in using practical field instruction to train his students. Werner's philosophy was expressed as follows: "Our earth is a child of time and has been built up gradually." His own theory of the origin of the earth was that all of the materials that now constitute the earth's crust were deposited successively from the waters of a large primeval ocean that at one time covered the whole earth and was higher than even the tallest mountains. Although Werner was aware that certain rocks and formations originated in the action of fire (for example, through the action of volcanoes), he held that this was minor compared to the major formation of successive layers of the earth's crust from material held in suspension or solution in the great primeval oceans. Because his theory was based on the action of water, it came to be known as Neptunian, and his followers were called Neptunists.

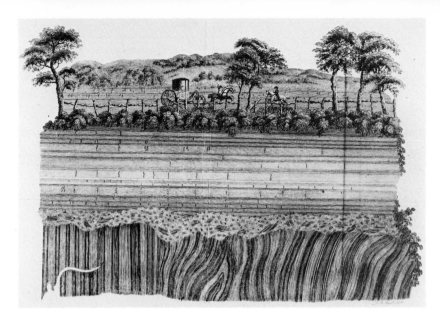

207. The Plutonian theory. In 1795 James Hutton published the first volume of his *Theory of the Earth, with Proofs and Illustrations.* The central theme of Hutton's work is that the key to the past is to be found in the present and that, accordingly, every aspect of the past history of the earth must be related to processes that we actually see going on all the time, or to processes that have been going on for a long period of time. Hutton was impressed by the fact that even sedimentary strata, which he presumed to have been laid down horizontally, were often found inclined, or even end on end, folded over on one another, or ending abruptly in a manner that indicated great convulsions of the earth itself. He attributed the subterranean force responsible for such action to the earth's internal heat, which he associated with the activity of volcanoes. These he regarded as "spiracles to the subterranean furnace in order to prevent the unnecessary elevation of land, and fatal effects of earthquakes." Such heat implied that the earth's nucleus might "be a fluid mass, melted, but unchanged by the action of heat"; accordingly, his theory became known as Plutonian. This picture from Hutton's book is a view of a region near Jedburgh, in the Southern Uplands of Scotland. Hutton's concept of the ever-continuing cycles of earth history, with "no sign of a beginning and no prospect of an end," essentially powered by the forces of heat, was symbolized by the living landscape and human activity atop the geologic structures.

208. Is it a crocodile? In 1799 Barthélemy Faujas de Saint-Fond described a gigantic reptilian skull, of which only the jaw was intact. It had been found in 1780 in the chalk deposits of a cave in the Mountain of Saint-Peter in Maastricht, Holland. Faujas's description of this find was printed in 1799 in a book on the natural history of the region that contains this illustration of the removal of the fossil jaw from the cave. The jaw remained in Holland from 1780 until 1795, when French troops sent it to the Muséum d'Histoire Naturelle in Paris. Faujas thought the skull belonged to a crocodile, but in 1824 Cuvier called it a "marine-like reptile." It is now considered to be an extinct form of lizard. This find led Cuvier to what he called a "most remarkable and most astounding result," that "the older the beds [of rock] in which . . . bones are found, the more they differ from those of animals we know today."

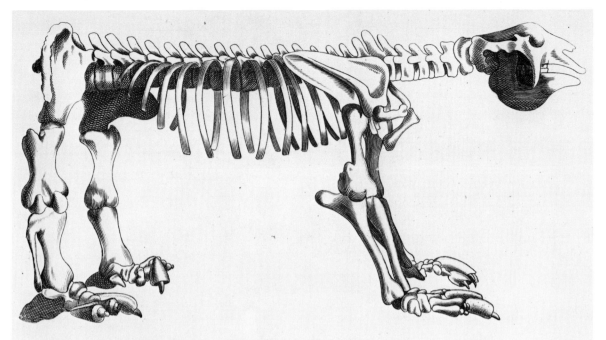

The SKELETON of a large species of QUADRUPED hitherto unknown *lately discovered one hundred feet under ground near the River la Plata.*

209. Skeleton of a megatherium. Cuvier, the founder of modern paleontology, invented the name *megatherium* (literally huge beast) and applied it to animals of great size, known to us only by their bones. The earliest drawing of a complete skeleton of the "huge fossil ground-sloth *megatherium* from Paraguay" was made by a Spanish artist. This illustration was reproduced from that drawing and published in the *Monthly Magazine and British Register* for September 1796. Through careful comparison, based on his extensive knowledge of comparative anatomy, Cuvier proved that this skeleton belonged to a species distinct from any known to be alive in recent historic times. In this way, he showed that species actually could become extinct.

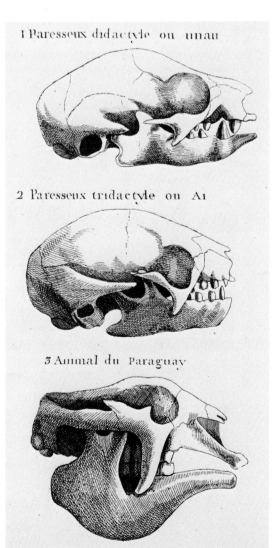

1 Paresseux didactyle ou unau

2 Paresseux tridactyle ou Aı

3 Animal du Paraguay

210. Cuvier applies comparative anatomy to paleontology. In a paper published in 1796, Cuvier compared the skulls of two species of living tree-sloths (1 and 2) with that of the giant fossil ground-sloth, or *megatherium*, from Paraguay (3), which, in the drawing, is reduced to the same size. The species difference is apparent. In another study, Cuvier made a detailed bone-by-bone comparison of "the skeletal anatomy of living and fossil elephants." His proof "that the Indian and African elephants were distinct species" has been described by Martin Rudwick as "a triumph for the new methods in anatomy, with their emphasis on dissection and internal features." Cuvier extended his argument to show that the Siberian mammoth, or fossil elephant, must have been "distinct from either of the living species."

211. Illustrating the fauna of New Guinea. This engraving made from a drawing by Pierre Sonnerat shows the artist at work, while a native holds a large leaf over his head to protect him from the sun. At the bottom of the picture can be seen other animals and specimens to be drawn for Sonnerat's report, *Voyage to New Guinea, in Which may Be Found . . . Details Relative to the Natural History of the Animal and Vegetable Kingdoms* (Paris, 1776).

18

New Worlds

One of the reasons why the exploration of new lands aroused so much excitement was the curiosity that arose concerning the newly discovered plants and animals and the nature and habits of the native men and women of these new lands. It was especially puzzling to discover that some of the animals in the New World were not identically the same as those in the Old World. Did this mean that the animals in the New World were part of an original creation that had been spared the Flood? If so, then the Flood would not have been universal. As one commentator put it, "I hold it un-Christian to say that the New World was spared the Flood." Or were the animals of the New World the result of a special creation after the Flood? If that was the case, then it could hardly be said that the creation had been "complete" in the traditional six days of the Scriptures.

Apart from the normal curiosity about the mineral resources and forms of life of newly discovered lands, there was a practical concern in learning whether these resources might be of economic benefit. For all these reasons, naturalists and explorers set out to write the natural history of the entire world and to describe all of its inhabitants.

In order to make certain that the specimens gathered by naturalists and explorers would arrive at museums and institutes in good condition, special sets of instructions were printed, which described ways of preparing skins of birds and animals, of packing seeds, and of mounting and preserving butterflies and other insects. Special devices were constructed to keep seedlings alive so that they could be planted in the gardens of Europe.

As time went on, American naturalists joined the search for plants and animals. One of them was John Bartram, whose house and garden may still be seen in Philadelphia. Wishing to make certain that the shrubs and plants he collected would arrive safely on the other side of the Atlantic, he attached special instructions with them to cover the contingencies of war. Should the British or American ship transporting the plants be captured by the French, he requested that his specimens be delivered to Georges Louis Leclerc de Buffon at the Royal Garden in Paris. In this way, some of the plants collected by Bartram ended up in France.

There were also "new worlds" to be found at home. Archaeological ruins, such as England's Stonehenge and the megaliths in Brittany, were vestiges of civilizations and cultures that required as much exploration as "new worlds" overseas. Antiquities were also being dug out of the ground, providing additional evidence of the cultural life and artistic pro-

ductions of a bygone age.

Plant exploration also received a considerable impetus from the development of formal gardens, particularly gardens laid out in the geometric style favored by the Dutch. There was a great desire by gardeners to have new and different ornamental plants, and these were introduced into Europe from all over the world. The rapid worldwide accumulation of plants produced a demand for a rational system of classification of plants, which was accomplished by Carolus Linnaeus in the eighteenth century.

212. The New World two decades after Columbus. In 1513 the Turkish sea captain Muhyi al-Din Piri Rais (or Re'is) drew a map of the South Atlantic, depicting the Iberian Peninsula, the eastern bulge of North America, the eastern coast of Central and South America, and some islands in the Caribbean. The map, drawn in color, contains pictures and notes concerning the countries of the New World, their human inhabitants, and the animals and plants to be found there. Apart from recording the information concerning the discoveries made by Christopher Columbus and other explorers, the map attracts attention because of its combination of imaginary beings and accurately rendered animals of the New World. This detail of the map showing South America includes a unicorn and a man "with his head beneath his shoulders," the same type of man described by Sir John Mandeville and thought to inhabit Central Asia.

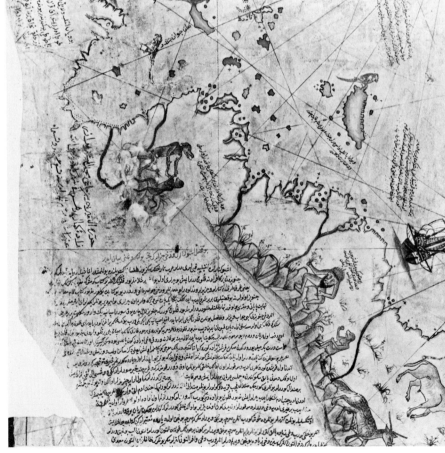

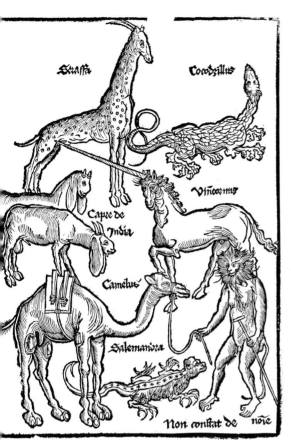

213. Animals encountered in the Holy Land. Bernhard von Breydenbach published an account of his pilgrimage to the Holy Land in 1486. Among his fellow pilgrims was the artist Erhard Reuwich, who illustrated Breydenbach's book with woodcuts. The animals are depicted, the artist assures us, "truly as we saw them in the Holy Land." Some of the animals are easily recognizable, such as the giraffe, crocodile, and dromedary camel (its first representation in a printed book), but the unicorn is pure fiction, as is the man-like ape holding a stick and leading the camel. Nor is it likely that in the Holy Land Reuwich would have seen an Indian rhinoceros, looking somewhat like a floppy-eared, double-horned water buffalo. The salamander at the bottom of the plate is also more fiction than fact.

214. Ancient ruins reveal the past. During the sixteenth, seventeenth, and eighteenth centuries, there was much interest in ruins, including Stonehenge, which appears in this engraving signed "R. F. 1575." In England such structures were often associated with the Druids and predated written records as clues to the past.

215. Indians in Virginia performing a religious dance. Among the masterpieces of accurate and artistic rendition of New World customs, and of the flora and fauna of America, is the outstanding series of watercolors by the Englishman John White. They were engraved by Theodore De Bry and first appeared in the second edition of Thomas Harriot's *Briefe and True Report of the New Found Land of Virginia* (1590), after which they were widely reproduced. Harriot described the scene as "a great and solemn feast," to which the Indians "come dressed in a strange fashion, wearing marks on their backs signifying the places they come from." They assemble around a series of "tall posts carved into faces resembling those of veiled nuns." At the center, "three of the most beautiful virgins, their arms about each other, turn around and around."

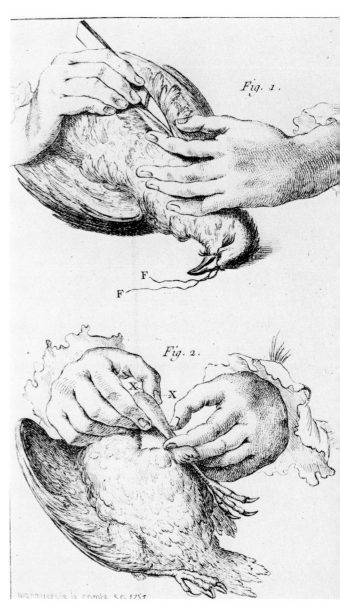

216. Georg Eberhard Rumphius, "Pliny of the Indies." One of the outstanding naturalists of the seventeenth century, Rumphius made the first systematic survey of the flora and fauna and the geographical and mineralogical features of the Moluccas, or Spice Islands. In 1670, barely past his thirtieth birthday, Rumphius became blind; but so great was his range of knowledge that he continued his work in natural history, depending on the eyes of assistants and his own sense of touch. The loss of his sight could not conquer his zeal for natural history.

217. Preparation of birds for preservation and shipping. Careful illustrated instructions such as these were given to those who were gathering plants and animals for the great collections in Europe to ensure that specimens would not deteriorate in transit. This instruction concerning birds appeared in a 1758 work published in Lyons, France, entitled *Instructive Memoir on the Method of Collecting, Preparing, Preserving, and Sending Different Curiosities of Natural History*. A supplementary memoir was entitled *Advice on the Shipment by Sea of Trees, of Living Plants, of Seeds. . . .*

164

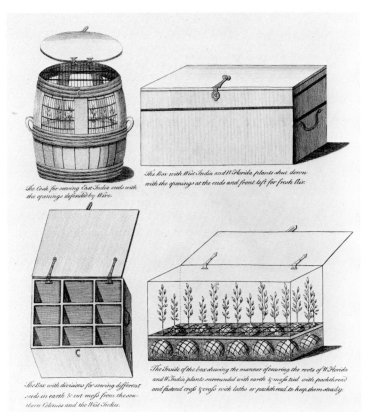

The Cask for sowing East-India seeds with the openings defended by Wire.

The Box with West India and W. Florida plants shut down with the openings at the ends and front left for fresh Air.

The Box with divisions for sowing different seeds in earth & cut moss from the southern Colonies and the West Indies.

The Inside of the box shewing the manner of securing the roots of W. Florida and W. India plants surrounded with earth & moss tied with packthread and fastend cross & cross with laths or packthread to keep them steady.

218. Methods of packing seeds and plants to be sent by ship. The seed and plant boxes shown here were described by John Ellis in 1770 in a book entitled *Directions for Bringing Over Seeds and Plants . . . in a State of Vegetation: Together With a Catalogue of Such Foreign Plants as Are Worthy of Being Encouraged in Our American Colonies for the Purposes of Medicine, Agriculture, and Commerce.*

219. Exchanging a handkerchief for a crayfish. This drawing was made by a member of the crew of Captain James Cook's *Endeavour* during the voyage to New Zealand in 1769–70.

220. Maoris expressing defiance. Reaching beyond the typical posed portrait, Sydney Parkinson recorded the New Zealand natives as they expressed their roles in society. Here he shows how "warriors defy their enemies." The drawing was published as one of the illustrations in Parkinson's *A Journal of a Voyage to the South Seas in His Majesty's Ship the Endeavour* (1773).

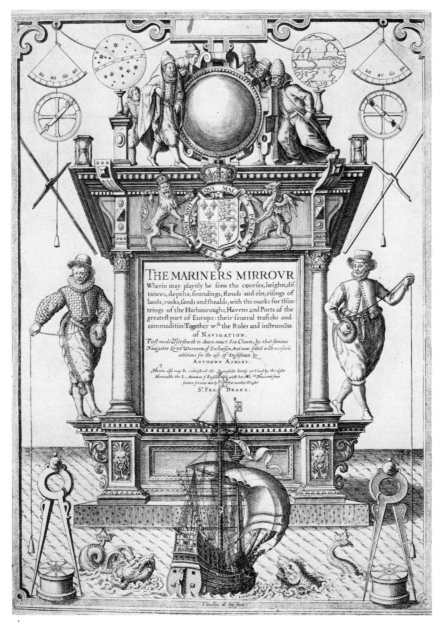

221. _The Mariners Mirrour._ Published around 1588, _The Mariners Mirrour_ was an English translation of a Dutch work on practical navigation of the same period. The title page shown here is almost an exact copy of the one in the original Dutch edition, except that the costumes of the navigators and mariners were changed to English dress. A pair of dividers for measuring distances on maps is placed in each of the lower corners. At center, left and right, sailors are using a sounding line to find the depth of water; and above each of them, from bottom to top, are three tools for astronomical navigation: a cross-staff, or Jacob's staff; a circular sea astrolabe; and a mariner's quadrant. A group of navigators and mariners at the top are peering into the "Mirrour of the Sea." To their left is an astronomical globe; to the right, a terrestrial globe. The hourglasses were used for measuring time, as in "dead reckoning." The astrolabe was used (often in conjunction with a nocturnal) to determine the altitude of the celestial pole, which equals the observer's latitude.

19

Mapping and Measuring the Earth

Although the size of the earth had been determined by Eratosthenes in ancient times, its exact dimensions were not generally known even in Newton's time, in the 1660s. The first accurate, modern determination of the earth's radius was undertaken late in the decade by the French astronomer Jean Picard, who was associated with the new Royal Academy of Sciences in Paris.

In the middle of the eighteenth century, another Frenchman, Pierre Louis Moreau de Maupertuis, made a different type of earth measurement. Maupertuis went to Lapland to find out how the length of a minute of arc along a meridian of longitude would compare to a similar distance nearer the equator. The result would prove whether the earth is oblate (flattened at the poles and bulging at the equator) or prolate (stretched out at the poles). The oblate shape had been predicted by Newton in his *Principia*, as part of his explanation of the phenomenon of precession, but the prolate shape was espoused by Gian Domenico Cassini, the director of the Paris

Observatory. Maupertuis's measurements were a great triumph for the Newtonian theory.

In the sixteenth and seventeenth centuries, the establishment of overseas colonies and the development of world trade produced the need for a convenient and accurate method of determining longitude at sea. Thus, initial support in the study of astronomy came from a desire to improve the methods of navigation rather than from an interest in the nature of the universe, and that was the primary basis for the establishment of royal observatories at Greenwich and Paris. Galileo, who first discovered the satellite system of Jupiter, hoped that tables of these satellites might provide the basis for an accurate method for determining longitude.

Pressure to determine longitudes was so great that early in the eighteenth century the British Parliament established "a Publick Reward for such a Person or Persons as shall discover the Longitude at Sea." The prize was to be £10,000 for a means of determining a ship's longitude to within 1 degree of arc, but as much as £20,000 if the accuracy of the calculation was within 0.5 degrees of arc, or 30 minutes. The importance of the problem is reflected in the cash value of the prize, the highest reward equaling about $1,000,000 in today's money. To understand why this problem was so critical, it must be remembered that in 1691, four years after the first edition of Newton's *Principia*, seven British warships were so ignorant of their position in longitude that they were wrecked near Plymouth. Similarly, three years later, another British fleet met disaster by sailing head-on into Gibraltar. Likewise, in 1707, a squadron of the Royal Navy, believing itself to be in a safe position, ran aground on rocks off the Scilly Isles, with a loss of some 200 lives and four ships of the line.

In the eighteenth century, two methods were introduced for finding longitude at sea. One depended upon the motion of the moon, but the other was mechanical and required the use of an accurate portable clock (the marine chronometer). The modern pendulum clock was first conceived by Galileo, but the design of a complete clock with a pendulum and modern escapement was first published by Christiaan Huygens in 1673. (Robert Hooke also has some claim to priority in the invention.) Such pendulum clocks, however, were not of much use on shipboard. The first practical ship's chronometer was invented by John Harrison, and was tested on one of Captain James Cook's voyages to the South Seas. With an accurate clock on board, a navigator could determine his longitude by finding his local time—a relatively simple job—and figuring out the difference between his local time and the time at Greenwich, as determined from a shipboard chronometer.

Two great events of the eighteenth century that helped establish a means of determining the true scale of the solar system were the transits of Venus in 1761 and 1769. In a transit, Venus is seen to cross the face of the sun. To date, only five transits have ever been observed: in 1639, 1761, 1769, 1874, and 1882.

A transit enables astronomers to determine the exact distance from the earth to the sun, the fundamental yardstick of our solar system. As the year 1761 became imminent, scientists from many countries made plans to observe the transit and expeditions were sent to many distant lands. The same thing happened in 1769, and both experiences were the first large-scale instances on record of international cooperation for a scientific purpose.

Toward the end of the eighteenth century, a new determination of the exact size of the earth was made in France. This measurement was important for the metric system, established after the French Revolution. The meter, the basis of the system, was originally intended to be one ten-millionth of the distance measured along a meridian of the earth from either pole to the equator. It has since been redefined several times and is now specified in terms of the wavelength of the orange-red line in the spectrum of one of the isotopes of krypton (^{86}Kr).

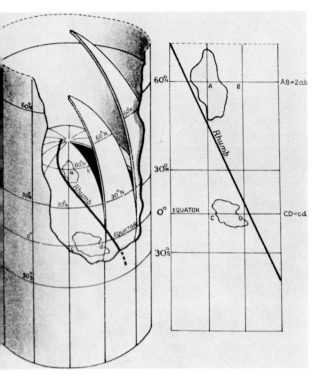

222. The Mercator projection. What we call a Mercator map projection is a means of projecting a spherical map (as of the earth's surface) onto a cylinder wrapped tightly around a sphere. When the cylinder is unrolled, a flat map will be obtained. A map based on this projection has a great advantage for navigators because a rhumb line (a line that is curved on a globe) will appear straight, as if drawn with a ruler. This line corresponds to the path of a ship sailing in a constant direction, that is, cutting across circles of longitude at a constant angle. Conceived in the sixteenth century by the Flemish cartographer Gerardus Mercator, this method of mapmaking was put into actual practice for the use of mariners in 1599 by Edward Wright, one of the great figures in the history of navigation. According to the historian of navigation David W. Waters, Wright was the first person to provide a chart upon which mariners "could plot the way of a ship through the sea . . . [yet] his greatest work—his chart projection—bears the name of another." Mercator's own chart was useful for geography, but could not serve the needs of navigators until redone by Wright. This diagram has been drawn by Waters to conform to Wright's description in his *Certaine Errors in Navigation,* first published in 1599, reprinted in 1610, and reported as "still being sold in 1694."

223. A chart of the world on Mercator's projection. This is an illustration of a pioneering sea chart based on the Mercator projection with the corrections and improvements introduced by Wright. Wright was a professor of mathematics at Caius College, Cambridge, and a colleague of John Napier, the inventor of logarithms. In Mercator's original projection, the circles of longitude and parallels of latitude appeared as a grid of straight lines with parallels of latitude and meridians of longitude drawn as straight lines at right angles to one another. As a result, shapes are distorted, particularly in the northern latitudes, and some distances are exaggerated. In order to use the projection, a set of tables was required to give mariners information concerning the actual distance. Mercator's own map was published on eighteen separate engraved sheets in August 1569. But it was not until Wright had made Mercator's pioneering work practical that the Mercator projection became the basis of the charts upon which every navigator came to depend. This chart comes from the third edition (1657) of Wright's *Certaine Errors in Navigation.*

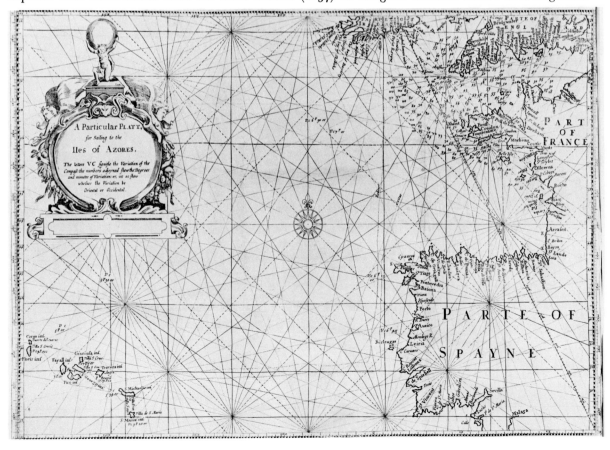

224. Picard's quadrant. In 1671 Jean Picard, a founding member of the Royal Academy of Sciences in Paris, undertook triangulation measurements to determine the length of a degree along a meridian. He was trying to find a way of determining the size of the earth and of increasing the accuracy of geographical coordinates on maps. It was Picard's value of the size of the earth that Newton used in his successful "moon test" of the inverse-square law of gravitation. This illustration clarifies the structure of Picard's quadrant as well as the way it was used. Picard made many improvements in the instruments used for geodesy, notably by adapting the astronomical telescopes to earlier instruments with open sights. Working with Adrien Auzout, Picard also improved the movable-wire micrometer, which he used to measure the diameters of the sun, the moon, and the planets.

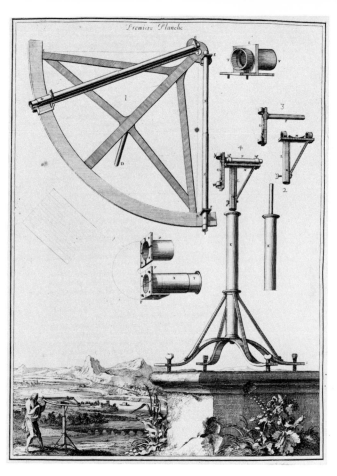

225. Halley's map of the Atlantic. This is the first map ever published showing "lines of equal magnetic variation." Edmond Halley hoped that the phenomenon of magnetic variation might serve as a basis for determining longitude, but this method did not prove effective. The map, now in the British Royal Geographical Society, has the rare printed instructions for its use attached to the edges. Even if magnetism did not provide a means of determining the longitude, information concerning the variation was of great use to navigators, since it enabled them to make corrections for discrepancies between the direction in which their compass needles pointed and the direction of true magnetic north. Halley pointed out that his chart was based on Mercator's projection, which "from its particular Use in Navigation, ought rather to be named the Nautical; as being the only True and Sufficient Chart for the Sea."

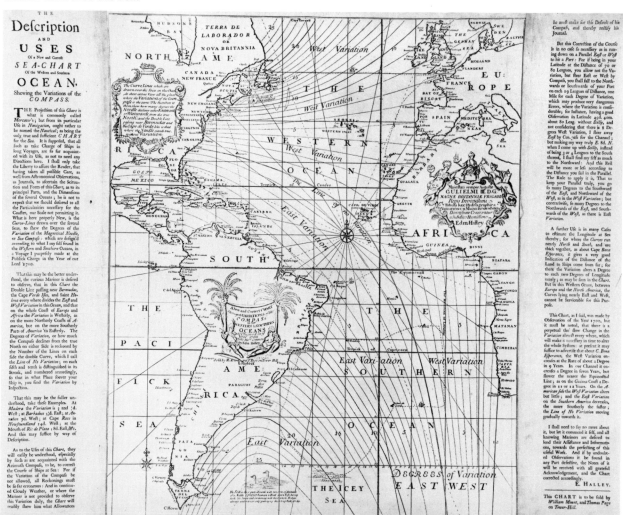

t Srade and ... *Divine Service*

226. Using a sextant. This rare illustration of the early use of a sextant appears in the manuscript log of the *Owen Glendower*, a ship of the late eighteenth century. In typical fashion, the sailor stands facing the sun with legs placed firmly apart while he holds the sextant at arm's length and looks at the reflected image of the sun in a mirror located on one arm of the instrument. A series of filters prevents blinding while "shooting" the sun. The mirror is moved until the reflected image of the sun coincides with the horizon line. The angle of the sun from the horizon can then be read on a scale on the sextant. Several readings are taken around noon to determine the maximum angle of elevation of the sun, which occurs when the sun (at noon) crosses the meridian; this value is added or subtracted to a number found for that time in a table of solar declinations, and the result is the latitude of the ship. At night, the sextant could be similarly used to measure the altitude of the North Star, which gives the latitude directly, but the value of such sightings was secondary to those of the sun.

227. The first practical marine chronometer. In the middle of the eighteenth century, John Harrison, a London clockmaker, succeeded in making a chronometer that would keep Greenwich time accurately enough to serve the needs of navigators. To determine longitude, the navigator had to find the difference between the local time at his position, which was moderately easy to determine by celestial observation, and the local time at some standard meridian, say Greenwich. Shown here is Harrison's first chronometer, which was completed in 1735 and taken on a voyage to Lisbon and back home in 1737. Obviously, it was somewhat cumbersome, but in 1761, Harrison produced a chronometer the shape and size of a large watch. This instrument was tested on a voyage to Madeira and Jamaica. During the 147 days of the trip the total error was 1 minute 54½ seconds, and that degree of accuracy was increased when the chronometer was later hung in gimbals, instead of being fixed to the ship.

Part Six

THE
LIFE
SCIENCES

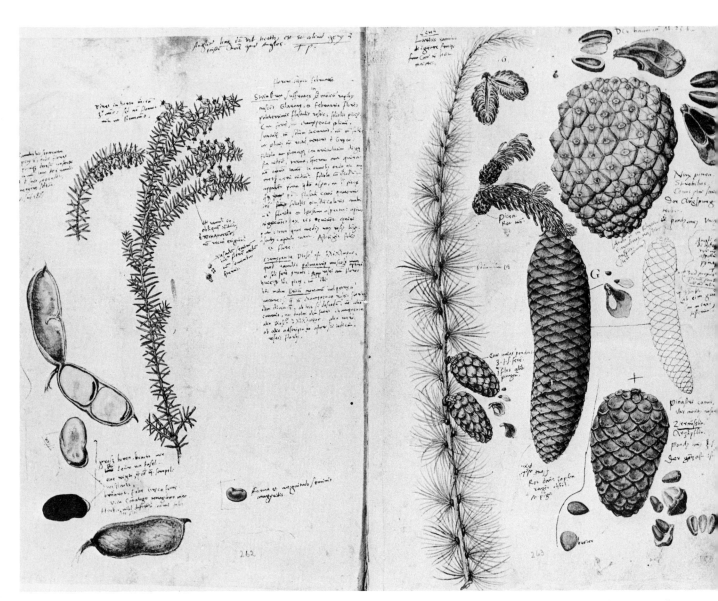

228. A naturalist's record. These pages from a notebook of the sixteenth-century Swiss naturalist Konrad Gesner contain carefully executed life-like drawings that form a lasting record of a naturalist's observations. They were also the basis of woodblocks that served as book illustrations. The right-hand sheet depicts the larch, the spruce, and the pine, which were traditional sources of pharmaceutical materials.

20

The World of Plants

During the sixteenth and seventeenth centuries, many new plants were added to scientists' lists of flora. With the introduction of the microscope in the seventeenth century, the cellular structure of plants became known, although no one understood the function of cells in either plants or animals. Thus, descriptions of cellular structures provided an interesting but not particularly significant addition to the body of knowledge about plants.

In the early eighteenth century, real progress was made in relation to plant physiology and reproduction. In plant physiology, the new advances were almost completely the creation of one British clergyman and amateur naturalist, Stephen Hales. Hales not only measured the relative rates of growth of different parts of stems and leaves, but also studied the ways in which root pressure forces sap upward through the stems and stalks of living plants. He also investigated the transpiration of moisture through the leaves of green plants in relation to the ascent of sap. In a

series of crucial experiments, he proved that during the process of growth, the solid matter constituting a plant's increase in weight comes in large measure from the air. Although there was considerable interest in Hales's experiments, especially those concerning the chemistry of the air, rapid advances in plant physiology were not made again until well into the nineteenth century. (Hales also studied the physics of the circulatory system and was the first person to measure blood pressure in animals.)

The other great advance in plant science during this period was the discovery that plants reproduce sexually. Strong evidence to favor this view came from experiments in controlled hybridization, and some of the earliest experiments and observations were reported from the colonies in British North America. The discovery of sexual reproduction in plants was important not only because it provided knowledge concerning a little understood aspect of plant life, but also because it formed the basis of the Linnaean system of plant classification. Linnaeus classified plants by the way they propagate their kind from generation to generation. This system made it relatively easy for anyone to identify a plant, particularly if a flower containing the sexual organs was available. The "lower" plants, those undifferentiated by flowers, were considered a wholly separate class with their own system of classification. Later on, a "natural" system of classifying plants, in which plants are grouped in natural families rather than arbitrarily according to the number and arrangement of the reproductive organs, replaced the Linnaean system. The Linnaean system then appeared to be artificial, since it put plants of very different types into a single category, a factor recognized by Linnaeus himself, who hoped to replace the sexual system with a "natural method."

A notable development in plant science was the rise of an extraordinarily high level of artistry in scientific illustration. Today it is difficult to decide whether to admire the work of a man such as Georg Ehret more for the accuracy or the consummate artistry of his illustrations. An outstanding presentation of plant studies was prepared by Robert John Thornton, in *A New Illustration of the Sexual System of Linnaeus*, which was published in sections between 1799 and 1807. Its supplement, *The Temple of Flora*, is one of the most magnificent publications ever devoted to plants, and contains colored mezzotints of each plant in its natural habitat on a scale and in a style never before attempted and possibly never again equaled.

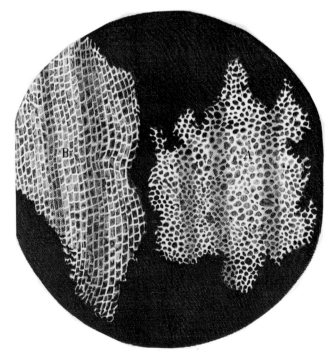

229. The first representation of cells. Robert Hooke's *Micrographia* (1665) contains a description and a drawing of the structure of cork as seen under the microscope. Hooke compared the construction he saw to the cells of a honeycomb. Actually, he was not observing cells as such, but rather the thickened walls of dead cells. In his drawings of other plants, however, for example the lower stomata of a stinging nettle leaf, it is possible to discern the outlines of cells. Leeuwenhoek also made drawings in which cells can be seen in living tissues, though neither he nor Hooke had any conception of the nature of the cell as it is understood today. In this drawing, the two light areas represent sections of a cork board made at right angles to each other.

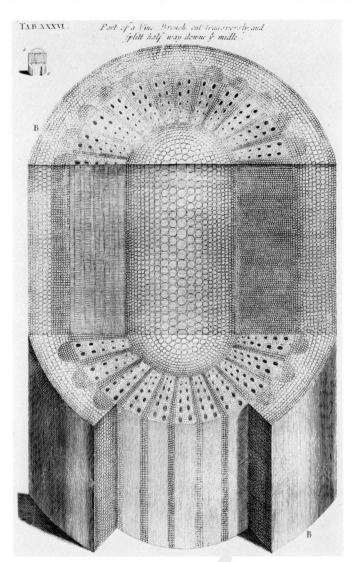

230. Section of a vine branch. In his *Anatomy of Plants* (1682), Nehemiah Grew included many illustrations of branches cut transversally, as well as vertically, to show their microscopic structure. He, too, was obviously familiar with the appearance of cells, although there is no indication of this concept in his writings. Grew had the insight to recognize flowers as "the sexual organs of plants."

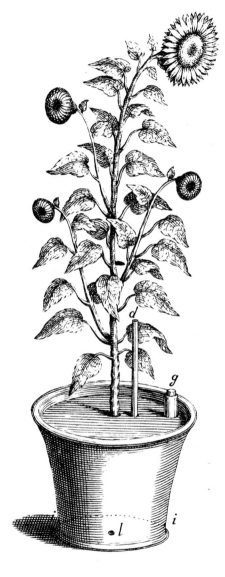

231. Hales's experiment on plant nutrition. Stephen Hales, often thought of as the father of plant physiology, attempted to find out where the matter of plants comes from. He grew a sunflower in a flower pot and covered the top of the pot with a sheet of lead perforated with holes for the stem of the plant and tubes through which he could pour measured amounts of water. Controlling all the factors, such as the rate of transpiration through the leaves, Hales was able to show that the increase in matter of the growing plant must come in large measure from the air.

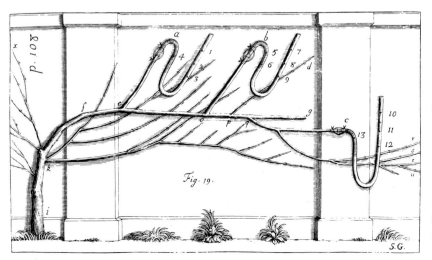

Fig. 19.

232. Measuring the pressure of the sap in vines. Hales called his 1727 study of plants *Vegetable Staticks* because he applied quantitative principles of weight and measure to the processes of plant life. Discovering that tips of vines will "bleed" if they are cut early in the spring, Hales made measurements of the sap pressure by fixing openended manometers (pressure gauges) filled with mercury to the cut ends of a vine.

233. Portrait of Linnaeus in Lapp dress. This painting of Linnaeus wearing the clothing of a Lapp huntsman and holding a shaman's drum in his left hand is one of the earliest known portraits of him. There are several versions of it, all painted by Martin Hoffman; it was also engraved as a mezzotint. A contemporary description of the portrait describes Linnaeus as presenting "the most grotesque appearance," with "boots of reindeer skin," having "about his body a girdle, from which was suspended a Laplander's drum, a needle to make nets, a straw snuff-box, a cartridge box, and a knife." In his right hand, Linnaeus holds a plant that he found in Lapland and that appeared to be "red from within and white from without." It is named after him, *Linnaea borealis.* The description of this portrait also reported that it "did not bear the least resemblance to Linnaeus in his age and maturity of manhood, except the piercing hazel eye, and the wart on the right cheek."

234. Flora Lapponica (1737). Linnaeus, the greatest botanist of the eighteenth century and one of the immortals of science, first achieved fame through his report on an expedition to study the flora of Lapland. Linnaeus not only described the flora of this then-unknown region of the north, but also reported on the habits and customs of the Lapps and the manner in which their lives were regulated by the migrations of reindeer.

235. Sexual reproduction of plants. Shown here are the original title page and frontispiece of Linnaeus's manuscript, *Praeludia Sponsaliorum Plantarum (Preludes to the Nuptials of Plants).* Written in 1729 in Uppsala, this work explains the structures of flowers in terms of their male and female characteristics. The number and arrangement of the sexual organs provided the basis for Linnaeus's system of classifying plants so they could be identified with ease. The manuscript was eventually published in Stockholm in 1746.

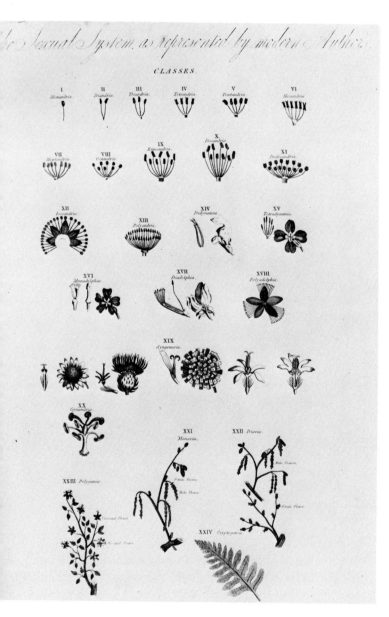

CLASSES.

I	II	III	IV	V	VI
Monandria.	Diandria.	Triandria.	Tetrandria.	Pentandria.	Hexandria.

236. The sexual system of Linnaeus. In 1799-1807 Robert John Thornton published *A New Illustration of the Sexual System of Linnaeus.* It was issued in segments and was supplemented in 1812 by *The Temple of Flora,* one of the most beautiful plant books ever published, containing many large-sized colored folios of plants in their natural settings. He illustrated some of the major aspects of the Linnaean system, in which flowering plants are divided into "classes" according to the number and arrangement of their anthers or stamens. The Linnaean system includes twenty-four primary classes, of which twenty-three are flowering plants, beginning with classes of plants that have both stamens and pistils in the same flower. Each of the classes was named for the number of stamens in the flower. Thus, *Monandria* (**I**) has one stamen; *Diandria* (**II**), two; up to *Dodecandria* (**XI**), any number between twelve and nineteen. There are two classes (**XII, XIII**) for flowers with twenty or more stamens. Two other classes of flowers (**XIV, XV**) have "stamens of markedly unequal length," completing the classes of plants with stamens "not united either above or below." Next, are five classes (**XVI** to **XX**) of plants whose flowers have "stamens united," then three additional classes (**XXI** to **XXIII**) comprising plants with "stamens and pistils in different flowers," and finally, the *Cryptogamia*—plants with hidden sexual organs, or "plants without proper flowers," such as mosses and ferns. All of these classes were then subdivided into orders according to the number and arrangement of the female organs. For example, the crocus belongs to the class *Triandria* and the order *Monogynia* (three stamens and one style); the lily and the snowdrop are both *Hexandria Monogynia* having six stamens and one style).

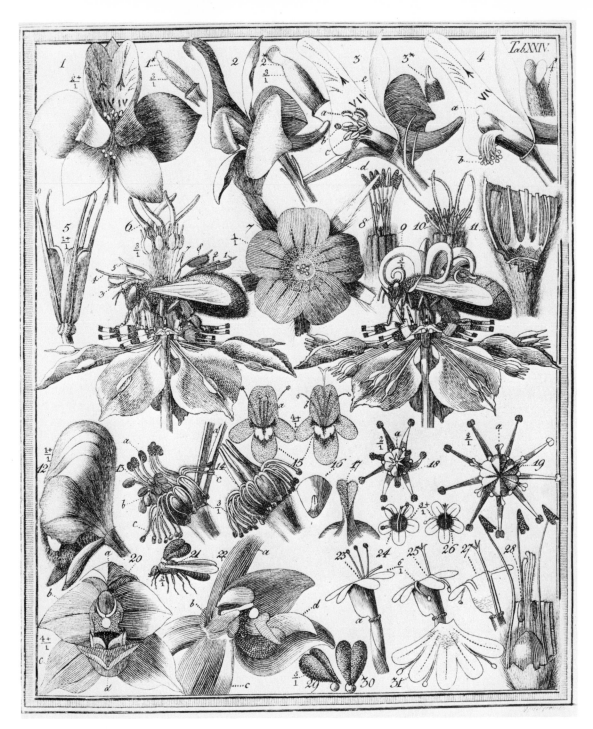

237. Insects and flowers. In 1793 Christian Conrad Sprengel published the results of careful studies detailing the relation of insects to plants, which he conceived as illustrative of the wisdom of the Creator. He believed that the function of the bright colors of flowers is to attract insects and that the purpose of insect visits to flowers is to carry pollen from stamens to pistils. Sprengel found that there are flowers that are invariably or almost always pollinated by the same insect, and he studied the way the size and shape of a given insect made it particularly suitable for pollinating a given plant. He compared the structure of flowers pollinated by insects and of those pollinated by the wind. He also showed that many bisexual flowers have stamens and pistils that mature at different times, making it impossible for the flower to be fertilized by its own pollen and thus requiring the action of insects or the wind for this purpose. He named this aspect of nature dichogamy, and concluded: "Nature does not appear to desire that a flower be fertilized by its own pollen."

Mandragora mas.
Mandragore.

238. Mandragora. In the late seventeenth century, the Paris Academy of Sciences planned an illustrated compendium of plant science that included Abraham Bosse's engraving of the mandrake, a plant noted for having a root that sometimes seemed to resemble the human form. Most of the other drawings in the volume were made by Nicolas Robert.

239. A flowering Sarasena. The artistic skill of William Bartram, the eighteenth-century American naturalist, is amply shown in his rendering of a Sarasena (*Sarracenia*) plant together with a large land snail, and a snake swallowing a frog. Bartram's father, John Bartram, was an outstanding plant explorer; he had such a gift for finding important new plants that Linnaeus is said to have called him the "greatest living natural botanist."

Duke of Cumberland

240. A Duke of Cumberland primrose by Georg Ehret (1740). The genius of Georg Ehret, according to historian Wilfrid Blunt, "was the dominant influence in botanical art during the middle years of the eighteenth century." Ehret was born in Heidelberg; worked in France, Germany, England, and Holland; and spent his mature years in London. He furnished botanical drawings for the engravings of some of the major books on botany and natural history and engraved some of the plates himself. His large-scale treatments not only revealed a plant in detail, but made an exquisite impression, exemplified by this striking watercolor of a primrose. Ehret wrote that "Linnaeus and I were the best of friends," and that Linnaeus showed him "his new method of examining the stamens, which I easily understood." Ehret decided to bring out a tabella (or picture) of it, and said he earned quite a bit of money from this tabella. When it was reprinted in Linnaeus's *Genera Plantarum* (1737) without any mention of Ehret as author, Ehret wrote that Linnaeus "appropriated everything for himself which he heard of, in order to make himself famous."

241. *Turnera ulmifloria*. Linnaeus's inventory and description of the plants in George Clifford's botanical garden in Haarlem is one of the truly great botanical books of all time. It combines the work of two young men of genius, both at the start of their careers, Linnaeus the botanist and Ehret the artist. The drawings by Ehret were engraved for this book by Jan Wandelaar, himself an extraordinarily gifted artist. It is this page of Linnaeus's book that appears in the eighteenth-century painting (reproduced in Chapter 2) by Jacob de Wit, in which two aged botanists and a young woman are shown admiring Linnaeus's *Hortus Cliffortianus*.

TAB X.

TURNERA e petiolo florens, foliis serratis. *Hort. Cliff.* 112. *fp* 1.
1. Ramus.
2. Folium ad cujus basin duæ glandulæ. Pedunculus e petiolo enatus cum calyce fructus, semine, stylis, stigmatibus.

G. D. EHRET del.   J. WANDELAAR fecit.

242. Thornton's _The Temple of Flora._ Accompanying Thornton's _A New Illustration of the Sexual System of Linnaeus_ was a series of beautifully colored prints comprising _The Temple of Flora. The Temple of Flora_ was published in 1812, and no cost was spared to make it the most sumptuous botanical publication ever produced. Thornton's aim was to show each plant in its natural setting. Of this "Large Flowering Sensitive Plant" Thornton wrote: "In the _large-flowering Mimosa (Callindra grandiflora_), first discovered in the mountains of Jamaica, you have the humming birds of that country, and one of the aborigines struck with astonishment at the peculiarities of the plant."

243. Anatomy of a horse. This skeleton of a horse, suspended by cords in a lifelike position, and with the hairs of the tail in place, is portrayed in a manner characteristic of the eighteenth century. The rendering is accurate and there is a scale (drawn between the legs), so that the actual size may be envisaged. The pedestal consists of some weathered carved stone blocks, to give a sense of the "antique," accentuated by the broken beam on which foliage has grown and which is holding up the skeleton itself. The naturalistic environment suggests the world of living nature, of which the horse is a part. Many similar backgrounds occur in French works of art of this period, in which there may be a beautiful lady on a swing rather than a horse's skeleton suspended from a beam or from a limb of a tree. This illustration comes from Philippe Étienne La Fosse's *Cours d'hippiatrique* (1772), the full title of which is (in translation) *Course on Hippiatry, or a Complete Treatise on the Medicine of Horses.*

21

The World of Animals

In the seventeenth and eighteenth centuries, knowledge concerning animals was extended from the macroscopic to the microscopic level, as it was in the world of plant science. Gross dissection was supplemented by a careful study of tiny animals and of the details of structures in animals both large and small that could be revealed only by the microscope.

From the sixteenth century on, the best books on animals and on natural history aimed to provide exact descriptions and representations of insects, birds, mammals, reptiles, and fishes. Such presentations were not confined to the fauna of the European countries in which their authors resided, but came to embrace the variety of animal life found in North America and South America, the East Indies and West Indies, Africa, and Asia. The authors were not content merely to give descriptions of external appearances, but also discussed the habitats of animals, their modes of hunting or food gathering, their sensory organs, their physiology, and even their reproductive systems.

Special attention was paid to the problems of the relationship of form to function. This reflected the emergence of a mechanistic view of nature and of the life processes.

Many books were written on the anatomy and physiology of the horse, in particular. This is hardly surprising in view of the great importance of horses as draft animals on farms, their use in hauling large loads—and their use as the chief means of transportation, with riders on horseback or in carriages or diligences. The high esteem in which the horse was held was expressed in a statement from Buffon's *Natural History*: "The horse is the noblest conquest made by man."

DES OYSEAVX, PAR P. BELON.
La comparaison du fufdit portraict des os humains monftre com
bien ceftuy cy qui eft d'vn oyfeau, en eft prochain.

Portraict de l'amas des os humains, mis en comparaison de l'anatomie de ceux des oyseaux, faisant que les lettres d'icelle se raporteront à ceste cy, pour faire apparoistre combien l'affinité est grande des vns aux autres.

Portraict des os de l'oyseau.

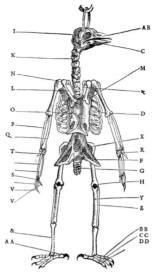

A B Les Oyseaux n'ont dents ne leures, mais ont le bec tranchant fort ou foible, plus ou moins selon l'affaire qu'ils ont eu à mettre en pieces ce dont ils vuent.
M Deux palerons longs & estroicts, vn en chascun costé.
҂ L'os qu'on nommé la Lunette ou Fourchette n'est trouvé en aucun autre animal, hors mis en l'oyseau.

D Six costes, attachees au coffre de l'estomach par devãt, & aux six vertebres du dos par derriere.
F Les deux os des hanches sont longs, car il n'y a aucunes vertebres au dessoubs des costes.
G Six osselets au cropion.
H La rouelle du genoil.
I Les sutures du test n'apparoissent gueres sinon qu'il soit boully.
k Douze vertebres au col, & six au dos.

d iii

244. Comparing the skeleton of a man and a bird. In 1555 Pierre Belon published *The History of the Nature of Birds*, which stressed comparative anatomy. By coding similar bones in different skeletons with the same letter, as shown here, he indicated their homologies. Belon reported that he had dissected 200 different species of birds. "It is therefore not surprising," he wrote, "that I am able to describe and figure the bones of birds so accurately." Stressing the astonishing affinity between the skeletal structure of man and bird, he stated that this relationship had not even been suspected before his time.

245. Rondelet's natural history of fish (1558). Guillaume Rondelet based his work on extensive research along the French Mediterranean and Atlantic coasts. His drawings were transformed into accurate woodcuts by Georges Reverdy, who succeeded in conveying the quality of different fish, as well as their anatomical details. Although Rondelet worked chiefly on fish, he also made studies of other groups, including the mammals. He wrote in his *History of the Fishes* that he had "with great labor and at great cost produced a work in which you [the reader] will find many good things of profit and satisfaction to studious men, and worthy of their commendation." The woodcuts in Rondelet's book are said (by F. J. Cole) to have been "so carefully executed that it is possible to identify the genus and often the species which was being examined."

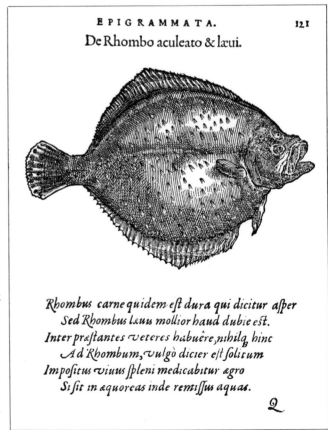

EPIGRAMMATA. 121
De Rhombo aculeato & læui.

Rhombus carne quidem est dura qui dicitur asper
Sed Rhombus læuis mollior haud dubie est.
Inter præstantes veteres habuêre, nihilq; hinc
Ad Rhombum, vulgò dicier est solitum
Impositus viuus spleni medicabitur ægro
Si sit in æquoreas inde remissus aquas.

𝒬

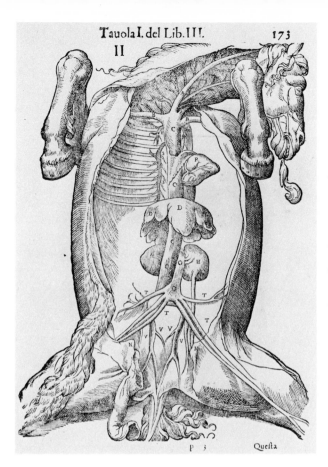

P 3   Questa

246. Heart and blood vessels of the abdomen and thorax of a horse. In 1598 Carlo Ruini published the treatise *Anatomy and Diseases of the Horse*. It has been described as "the first comprehensive monograph on the anatomy of an animal." Ruini's treatise was extremely popular, and between 1598 and 1769 it was printed in fifteen legitimate editions, plus a number of pirated versions. The style of the illustrations was adapted from the pictures used by Andreas Vesalius for human anatomy. Ruini presents his subject by beginning with the most internal parts of the horse, working outward by degrees to the surface musculature.

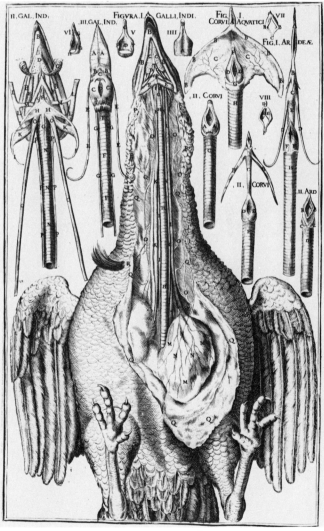

247. Larynx and associated muscles of a turkey. Giulio Casseri, a pupil of the great anatomist Hieronymus Fabricius of Aquapendente, was interested in the sensory organs and in the mechanism of the voice. He published accurate and detailed anatomical studies of the vocal organs in man, cats, rabbits, dogs, and birds, and he also studied the calf, horse, pig, sheep, ox, and goat. His book, from which this illustration is taken, contains discussions of the causes and production of sound in mammals and birds, and even in cicadas. Casseri not only described the production of sounds, but also devoted a considerable part of his book to hearing, paying special attention to the anatomy of the ear and the functions of its various parts.

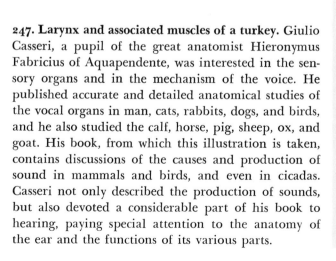

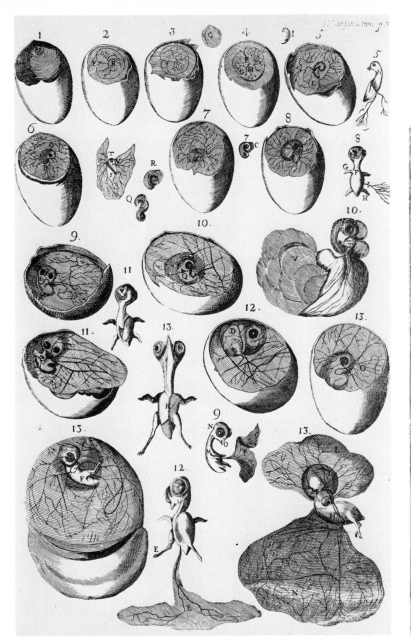

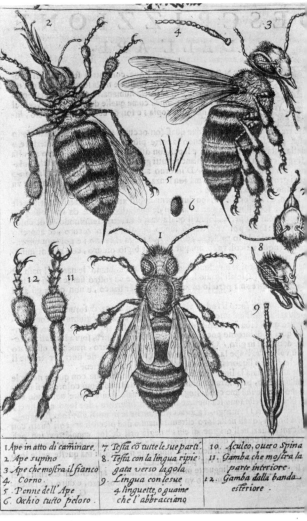

1 Ape in atto di caminare.	7 Testa cō tutte le sue parti.
2 Ape supino	8 Testa con la lingua ripie-
3 Ape che mostra il fianco	gata verso la gola
4 Corno.	9 Lingua con le sue
5 Penne dell' Ape	4 linguette, o guaine
6 Occhio tutto peloso	che l' abbracciano

10 Aculeo, ouero Spina
11 Gamba che mostra la
parte interiore.
12 Gamba dalla banda
esteriore.

248. Development of the chick embryo. Fabricius of Aquapendente was the teacher of William Harvey and the discoverer of valves in the veins. His treatise *On the Formation of Eggs and Chickens* (1621) is noted for its fine and accurate illustrations, such as this one. Although his descriptions show keen powers of observation, his theoretical discussions were wide of the mark and based on false inferences. For example, he held that both the yolk and albumen of the egg provide food for the developing embryo, because neither of them is present at the end of incubation.

249. Microscopic observations of the honeybee. Francesco Stelluti, a member of the Lincean Academy, of which Galileo was also a member, wrote in 1630 that he had "used the microscope to examine bees and all their parts. I have also figured separately all members that discovered to me, to my no less joy than marvel, since they are unknown to Aristotle and to every other naturalist." Stelluti said that "in compliment to our noble Lord Pope Urban VIII," he had caused to be "engraved here in Rome . . . three enlarged bees, drawn in such detail as was revealed by the glasses of the microscope and figured from front, back, and side. Stelluti originally had the figures printed on a single loose sheet; later they were re-engraved and printed Rome in 1630 to accompany an Italian translation of the poems of Persius. This illustration comes from the 1630 printing.

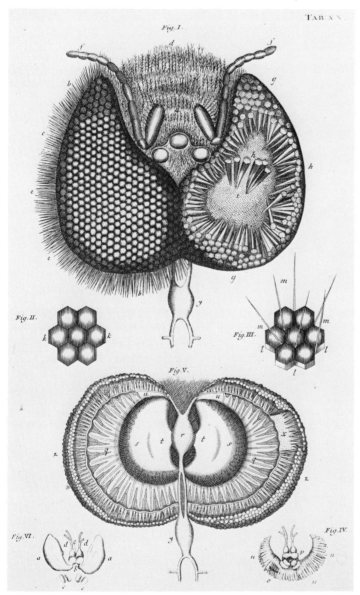

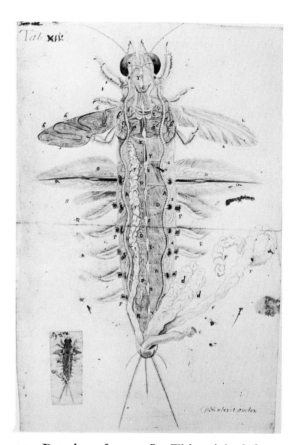

250. Compound eyes of a bee. Jan Swammerdam, a Dutch naturalist, is thought by some to be the greatest comparative anatomist of the seventeenth century. His study of the anatomy of the honeybee was the first comprehensive account of the insect, and its thoroughness is aptly demonstrated in this drawing of a bee's eye. In addition to illustrating the anatomy of the male, female, and drone, he explained the life history of the bee and the general economy of the hive. All of this material was published in his *Bible of Nature* (1737–39).

251. Drawing of a mayfly. This original drawing of the dissection of the male mayfly nymph was made as part of the preparation of Swammerdam's monograph *Ephemeri vita*, which was published in Amsterdam in 1675. Swammerdam's handwriting (*lower right*) reads *Delineavit auctor* ("drawn by the author"). In his book, Swammerdam espoused the doctrine of "preformation," according to which tiny preformed individuals are contained within the adult. A deeply religious man, Swammerdam sought a union of biology and religion, drawing morals for man from the study of insects. Thus he wrote that the study of ephemera may "give us wretched mortals a lively image of the shortness of this present life, and thereby induce us to aspire to a better one."

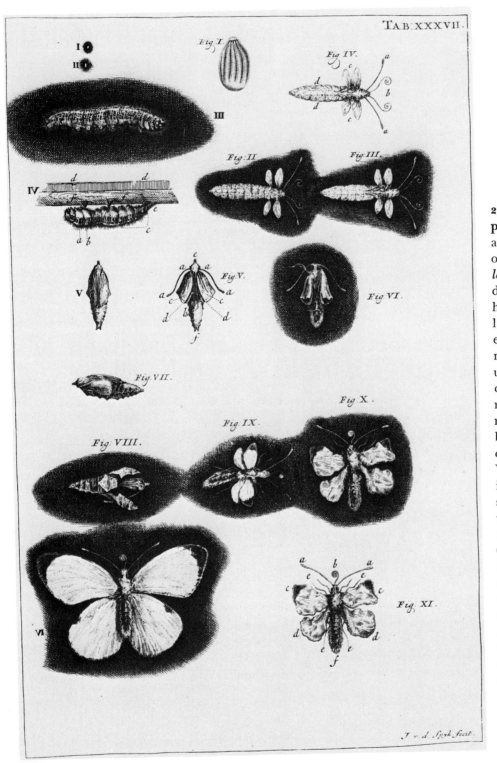

TAB. XXXVII.

J. v. d. Spyk fecit.

252. Lessons from the metamorphosis of insects. Swammerdam also illustrated the life history of a butterfly. Stage IV (*upper left*) is a caterpillar about to undergo metamorphosis. By using hot water to soften the external layers, Swammerdam was able to extract the butterfly from the metamorphosed caterpillar. (Figures II and III). A later stage of development, the chrysalis, is represented in Stage V. The rough shape of a butterfly can be seen when the chrysalis is opened up. Finally, in Stages VIII through XI, the escaping imago expands and reveals itself to be a perfect and fully developed butterfly. According to the anatomist and historian F. J. Cole, this evidence indicated to Swammerdam "that the butterfly was in the caterpillar, or in other words, the caterpillar is the butterfly and hence there is no epigenesis." There was at this time a debate as to whether organisms were preformed or developed gradually (epigenetically) from undifferentiated matter. Swammerdam's observations seemed to provide evidence in favor of preformation.

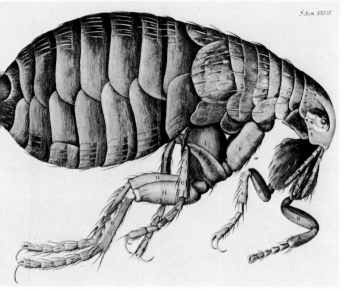

253. The "strength and beauty" of a flea. One of the most influential books in the history of microscopy was Robert Hooke's *Micrographia* (1665). In it, Hooke described his adventures in using the microscope to study organic and inorganic specimens. The *Micrographia* was especially notable for its plates, which were later reprinted separately. Some of them were foldout plates, and that of the flea was a spectacular 2 feet long. Hooke wrote that "The strength and beauty of this small creature, had it no other relation at all to man, would deserve a description. . . . The *Microscope* manifests it to be all over adorned with a curiously polished suit of *sable* Armour, neatly jointed, and beset with multitudes of sharp pinns, shaped almost like Porcupine's Quills, or bright conical Steel-bodkins."

254. Chick embryo seen through a microscope. For many years Marcello Malpighi used a powerful simple microscope to study the early stages of the development of the chick embryo. In 1673 he completed two short treatises, *On the Incubation of Eggs*, from which this illustration comes, and *On the Formation of the Chick in the Egg*. The latter was a landmark work, which has been said to have "pushed back the description of the embryo into the very first hours of incubation."

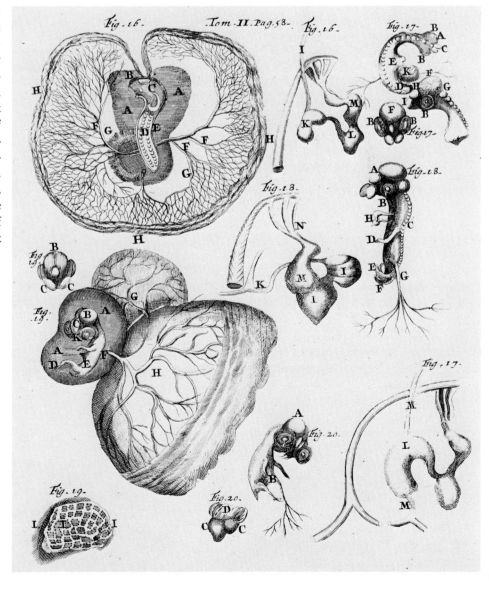

255. An "Orang-Outang," or chimpanzee. In his *Orang-Outang, sive Homo Sylvestris: or, the Anatomy of a Pygmie Compared With That of a Monkey, an Ape, and a Man* (1699), Edward Tyson described this figure of a chimpanzee as a "pygmie" or "homo sylvestris." His conclusion indicates the primitive state of knowledge of primate anatomy: "Our *Pygmie* is no *Man*, nor yet the *Common Ape*; but a sort of *Animal* between both." "Orang-Outang" was a Malaysian term meaning "man of the woods"; it appeared for the first time in a scientific book in 1641 written by Nicolaas Tulp who, like Tyson, used it in reference to the chimpanzee.

256. Humor in comparative anatomy. In his *Osteographia, or the Anatomy of the Bones* (1733), William Cheselden, who had been Isaac Newton's doctor, attempted to show the relationship of anatomy and structure to the habits of animals. His humorous skeletal representation of the encounter between a dog and a cat shows the character of each of these animals. The cat, a climber of trees, has back legs suitable for jumping and a flexible spinal column, which is arched in the drawing. The dog has strong legs, more suitable for a running animal, and it has a stiff spinal column since its body needs to be steady when moving swiftly. The expression in the skulls of these two animals is conveyed in a remarkably realistic manner.

257. The Hamburg seven-headed Hydra. A page from Albert Seba's *Thesaurus* (or *Treasury*) of natural history (1734) shows the American "flying dragon" (a kind of lizard) and two American birds. The stuffed specimen of a seven-headed monster was accepted by naturalists as if it had been a living creature until Linnaeus, in 1735, showed it to be a composite; the feet and jaws came from weasels and the body skin came from snakes. "Many people said it was the only one of its kind in the world," Linnaeus wrote, "and thanked God it had not multiplied." Linnaeus assumed that the Hydra had been put together by monks to illustrate one of the beasts described in the Apocalypse.

258. Corals at low tide. The frontispiece to John Ellis's *Natural History of Corallines* (1755) was captioned: "Groupes of different Corallines growing on Shells, supposed to make this Appearance on the Retreat of the Sea at a very Low Ebb Tide." Ellis, a British naturalist, was agent for West Florida in 1764 and for Dominica in 1770. His book, reprinted often and translated into French, established the fact that coral formations are collections of the skeletons of tiny animals despite their plantlike appearance.

259. Buffon. Georges Louis Leclerc, Comte de Buffon, began his career as a student of mathematics, working on problems of ship construction. He made a serious contribution to probability theory, studied geology, and translated Newton's *Method of Fluxions and Infinite Series* into French. From laboratory experiments on cooling, he estimated the age of the earth at 75,000 years, ten times older than was commonly believed. His ideas about the development of the earth were Plutonian rather than Neptunian, and he described a series of "epochs" during the last of which the earth's surface was modified by the action of man. Buffon also became administrative director of the Jardin du Roi in 1739, a post he held for fifty years. During that period, he wrote many volumes of his *Histoire naturelle*, a series of studies that set a new standard for books on the subject. In these books, Buffon's great literary style and the accuracy of his description was matched by the splendor of the printing and quality of the illustrations. This portrait was drawn by Louis Carmontelle in 1766.

260. Buffon and his colleagues engaged in research on generation. This original drawing by De Sève shows Buffon seated at the end of the table next to the Abbé John Turberville Needham, while Louis Daubenton looks through a microscope. At right, a surgeon has completed a dissection of a virgin bitch. This is the artistic rendering of the supposed discovery by Buffon and his colleagues of spermatozoa in the Graafian follicles of a virgin mammal: an impossible phenomenon that—according to Joseph Needham—must have been "owing to some mistake which has never been explained." Buffon was an ardent opponent of the theory of preformation. Instead, he embraced the principle of epigenesis and spent a great deal of time studying the development of embryos in animals.

261. Opening page of Buffon's *Histoire naturelle* (Paris, 1749). A vignette at the top of the page shows Buffon's conception of natural history as the union of the "natural" and "exact" sciences. Together with plants and animals of all sorts are pictured a telescope, an armillary sphere, and other symbols of the exact sciences. Note that one of the mathematical diagrams is the Pythagorean theorem. "Natural History," says Buffon, "embraces all the objects which are presented to us by the Universe . . . and offers to the curiosity of the human mind a vast spectacle, of which the totality is so great that it appears to be and actually is inexhaustible."

HISTOIRE NATURELLE.

PREMIER DISCOURS.

De la manière d'étudier & de traiter l'Histoire Naturelle.

L'HISTOIRE Naturelle prise dans toute son étendue, est une Histoire immense, elle embrasse tous les objets que nous présente l'Univers. Cette multitude prodigieuse de Quadrupèdes, d'Oiseaux, de Poissons, d'Insectes, de Plantes, de Minéraux, &c. offre à la curiosité de l'esprit humain un vaste spectacle, dont l'ensemble est si grand, qu'il paroît & qu'il est en effet inépuisable

A ij

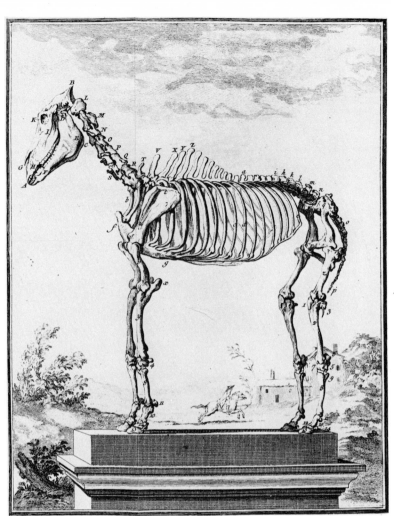

262. A skeleton of a horse. The remarkable aspect of this illustration from Buffon's *Histoire naturelle* is the placement of a skeleton on a pedestal in a rural landscape. A provocative contrast to the skeleton is made by the horse and rider in the background.

263. Anatomical drawing of a horse. This engraving was made by George Stubbs, one of the leading painters of animals of his time, for a large work entitled *Anatomy of the Horse* (1766). The engraving was accompanied by an outline drawing covered with tiny numbers and letters that could be used to identify the individual bones.

264. An anatomical lecture on the horse. The study of the anatomy and physiology of the horse was of practical as well as general scientific interest in the eighteenth century. All true gentlemen of that time were concerned for the health and well-being of their steeds and, accordingly, studied the structure of the horse to a degree that can only be understood in terms of the eighteenth-century style of life in the upper classes of society. The illustration shown here comes from Philippe Étienne La Fosse's *Cours d'hippiatrique* (1772).

265. Generation cycle of the frog. In 1785 Lazzaro Spallanzani published an account of his experiments on the generation of animals and plants. The opening plate, reproduced here, shows the frog's complete cycle of generation in a series of eight figures, beginning with the fertilized ovum at upper left and ending with the developed tadpole.

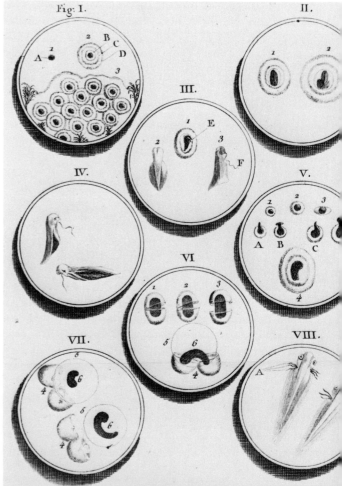

ARDEA Gularis

266. Form follows function. According to the French naturalist Jean Baptiste Lamarck, this shore bird from Senegal stretched itself and elongated its legs to keep its body dry. According to his theory that acquired characters are inherited, the bird's position produced anatomical changes over the centuries, which were then inherited. Eventually, the legs of the bird became like stilts. Lamarck began his work at the end of the eighteenth century, and from then on this doctrine of the inheritance of acquired characters was widely discussed, often in relation to a theory of evolution in animals.

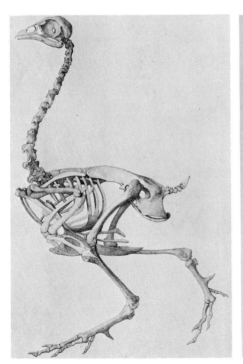

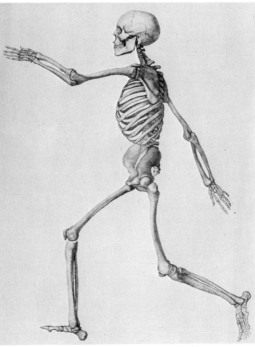

267. Comparative anatomy of man and fowl. From 1804 to 1806, the renowned painter George Stubbs completed a set of engravings intended for *A Comparative Anatomical Exhibition of the Human Body with That of a Tiger and a Common Fowl*. Some of the drawings were engraved by Stubbs, but the work was never completed. These two prints, now in the Royal Academy of Arts, London, compare the skeletons of a man and a fowl in running positions. It may be noted that the rib cage of the man is almost in contact with his pelvis; this inaccuracy suggests that the drawing may have been made very early in Stubbs's career.

268. Vesalius, founder of modern anatomy. This portrait is from Vesalius's treatise *De humani corporis fabrica (On the Fabric [or Construction] of the Human Body)* (1543), a large folio volume of 663 pages, illustrated with more than 200 spectacular woodcuts. Vesalius was twenty-eight when this portrait was made in 1542.

22

The Fabric of Man

From the fifteenth century on, dissections of human corpses, primarily those of criminals, provided knowledge for the scientist and the doctor and a kind of moral entertainment for curiosity seekers. Annual public anatomies attracted men and women from all classes of society, eager to see demonstrations of the composition and structure of the human body. Scenes of such anatomies, recorded by artists in many countries, illustrate the transition from a makeshift dissecting table, often set up in a chapel or other public building, to permanent anatomical theaters and museums. Eventually attendance at these dissections was required for medical students and even for practicing physicians.

Typically, the fifteenth-century dissection featured a reader who stood or sat apart on a high platform while two assistants performed the work. The reader usually had a massive tome before him, most often the work of the second-century physician Galen, but sometimes the writing of a medieval commentator on Galen, such as Mondino de' Luzzi. By the

sixteenth century, however, the reader was placed closer to the corpse and finally put the book aside and began to perform the anatomy himself, although often a book was still in evidence.

When not used for performing anatomies, the anatomical theater sometimes served as a cabinet of natural curiosities. On the payment of a suitable fee, ladies and gentlemen could visit the room to see the skeletons of human beings or other animals, various stuffed animals, and even a partly dissected cadaver. Since the corpses dissected were often those of criminals, it was common practice to flay such a body first and then tan and exhibit the skin as a lesson to others. A poem written on the occasion of the opening of the Amsterdam Anatomical Theater makes the point that "Evil doers who while living have done damage are of benefit after their death." Various parts of their bodies—kidney, tongue, heart, lung, brain, bone, hand—"afford a lesson to you, the Living." Skeletons in the museum or anatomical theater usually carried banners proclaiming the transitory nature of life and its fragility, thus emphasizing death as the counterpoint to life.

As detailed and accurate knowledge about the human body increased, great attention was paid to the skeletal system as well as to the structure of the muscles, two aspects of the body that were also important to painters and sculptors. While the anatomists served scientists, physicians, and artists alike, artists themselves helped science by making accurate and beautiful anatomical illustrations.

The great innovative work in human anatomy was Andreas Vesalius's treatise *De humani corporis fabrica (On the Fabric [or Construction] of the Human Body)* published in 1543, which stressed form in relation to function, or anatomy in relation to physiology. As scientific works of art, the magnificent plates drawn for Vesalius's book have probably never been surpassed. Unfortunately, the artists who did the drawings and cut the wood blocks remain anonymous, so attributions are at best conjectural.

In many eighteenth-century representations of the human body, the standard of art matched the level of knowledge. The books of Albrecht von Haller, Christoph Jacob Trew, Paolo Mascagni, and Jan Wandelaar demonstrate equally high levels of graphic skill and scientific knowledge.

Color printing was also attempted in the eighteenth century. A treatise on anatomy was planned by Jacques Christophe Le Blon, in which the plates were to be printed by means of a new color process. This work was never completed, however, and only one perfect example of a single plate is known today; it is in a private collection in Paris. In the absence of color, various schemes were invented to give the viewer an idea of the complexities in overlapping systems of organs and other structures. Leonardo invented "exploded" diagrams and a system of fine threadlike lines to indicate overlapping layers of muscles, while Vesalius, in successive illustrations, peeled open the human body layer by layer, down to the innermost muscles and skeletal structure.

As might be expected, progress in anatomy came to depend more and more on studies that focused on particular parts of the body. The new knowledge eventually demanded revisions of accepted physiological beliefs. The old physiology based on the Galenic concept of body "humors" was pushed aside and replaced by a new physiology based on more certain and detailed anatomical knowledge. In the seventeenth and eighteenth centuries, physiology became the physical and chemical study of bodily functions.

Although improved knowledge of the human body was useful for the physician and surgeon, it could not be put to use for internal surgery without anesthesia and antiseptics, which did not yet exist. Most surgeons therefore confined their attention to setting broken bones, amputating limbs, removing kidney stones, and treating wounds and injuries; for these tasks, detailed anatomical knowledge about the internal fabric of the human body was not of much practical use.

269. The old anatomy. This view of a dissection is from the 1491 edition of the *Fasciculus medicinae* of Joannes de Ketham, but it is almost an exact copy of the medieval manuscript versions. In typical medieval fashion, the professor, usually a medical doctor, dominates the scene as he reads a description from a book, presumably written by Galen. He is considerably removed from the actual dissection being performed by an assistant. In such representations, a professor of surgery usually stands beside the dissector, or prosector, to demonstrate the actual organs. In English hospitals and medical schools the academic title of "reader" is still used, as are those of "demonstrator in anatomy" and "prosector." It should be noted that there are not many students in attendance and that few of them are paying any attention to the demonstration.

270. The new anatomy. The title page from Vesalius's treatise of 1543 on the fabric of the human body exemplifies the techniques of the new science in graphic form. Instead of reading from one of the books of an ancient authority or relying on a demonstrator or a prosector, here the professor, Vesalius himself, is seen performing the dissection. Note that Vesalius is actually touching the cadaver with his hands, for, as he explained, anatomy had declined since Roman times because "physicians left manual work to others." The living human model holding on to the column (*left*) and the mounted skeleton (*center*) were both used to relate the organs being dissected or discussed to the whole body. In contrast to the preceding picture, the artist has caught the excitement and interest generated by a demonstration based on nature rather than on traditional learning. There is a large crowd paying close attention to the lecture and demonstration.

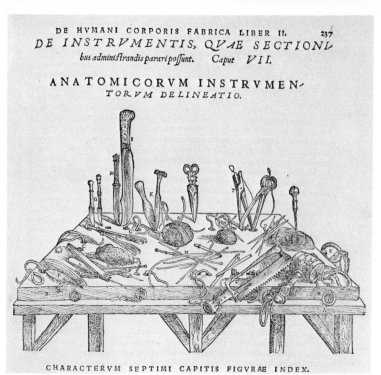

DE INSTRVMENTIS, QVAE SECTIONI-
bus administrandis parari possunt. Caput *VII.*

ANATOMICORVM INSTRVMEN-
TORVM DELINEATIO.

CHARACTERVM SEPTIMI CAPITIS FIGVRAE INDEX.

271. The tools of the anatomist. Vesalius's 1543 treatise shows not only the results of dissection, but the actual tools of the anatomist. Vesalius was thus indicating to his readers that they might do dissections for themselves, if supplied with the proper knives, saws, mallets, and so on. He was also emphasizing as dramatically as possible that the illustrations and text had been based upon actual dissections or actual examinations of nature, and not merely on the traditions of ancient and medieval learning.

272. Dr. Pieter Paaw performing an anatomy in Leiden in 1616. Public anatomies were often festive occasions, much like theatrical productions, for which tickets were sold. In this illustration, some foreigners are shown attending the anatomy, judging by their cassocks and hats. The banner held by the skeleton reads *Mors ultima linea rerum* ("Death is the last boundary [of human affairs]").

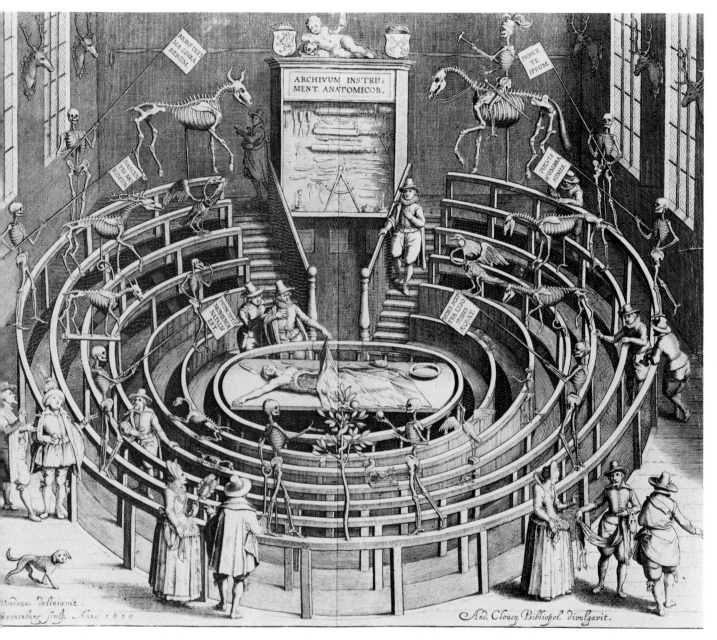

Inside the illustration:

ARCHIVUM INSTRU-
MENT. ANATOMICOR.

MORS ULTI-
MA LINEA
RERUM.

NOSCE-
TE
IPSUM.

NASCEN-
TES MO-
RIMUR.

PULVIS
UMBRA
SUMUS.

PRINCIPIUM
MORIENDI
NATALIS
EST.

MORS SCEP-
TRA LIGO-
NIBUS
AEQUAT.

Joudanus deliniavit.
Swanenburg sculp. Anno 1610

And. Cloucq Bibliopol. divulgavit.

273. The anatomical theater as a museum. Many early anatomical theaters, like this one in Leiden, doubled as museums of natural history when not in use for dissections; and dissected corpses were sometimes left in view for public inspection. Mounted on the walls are permanent displays of stuffed animal heads, and there are skeletons of human beings and animals here and there in the amphitheater. In the center foreground are skeletons representing Adam and Eve; Eve holds an apple in her skeletal right hand, and there is a stuffed snake in the tree of knowledge of good and evil. At the very back of the room (*upper right*), a skeletal knight rides a skeletal horse. Various human skeletons hold banners with moral lessons of life and death, including a quote from Horace: *Pulvis et umbra sumus* ("We are but dust and shadow").

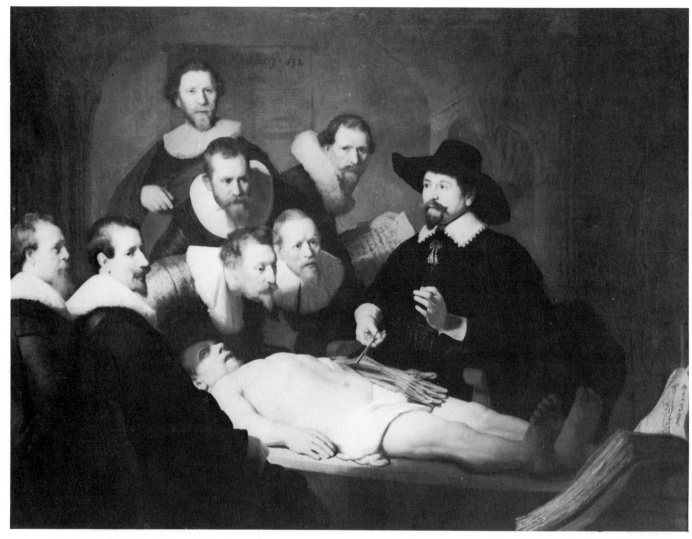

274. Rembrandt's *Anatomy Lesson of Dr. Tulp* (1632). At the age of twenty-six Rembrandt was commissioned to do a painting of Nicolaas Tulp performing an anatomy for members of the Guild of Surgeon-Anatomists of Amsterdam. The corpse is that of a young criminal, Aris Kimdt. Rembrandt's work is remarkable for its portrayal of the real involvement of the participants in the lesson and, as one historian has commented, "It hints at the possibilities of collective research," which later was to produce such remarkable results.

Anathomia oder abconterfettung eines
Weibs leib/wie er inwendig gestalret ist.

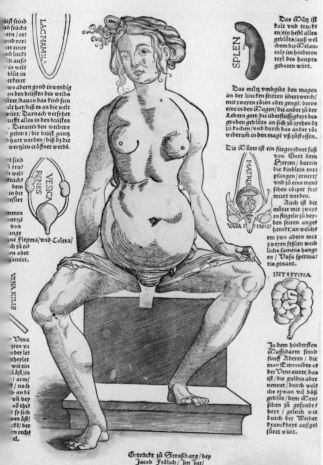
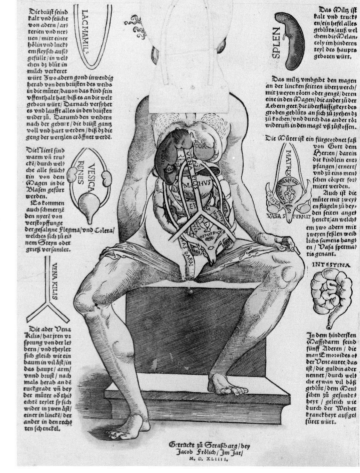

275. A female anatomical figure. Many enterprising printers of the sixteenth century used a novel way of showing the human abdomen and thorax in what has been called a "pseudo-three-dimensional form." This figure was published by Jakob Frölich in Strasbourg in 1544. The various layers revealed by an anatomy were printed on separate small sheets, which could be partly cut out and then glued together at the upper end, so as to form a series of flaps that could be turned back one by one "to reveal the successive stages of an anatomical dissection." These were widely copied and reprinted by various printers. In the male figure of this very rare pair of broadsides (not shown here), some parts of the text and even some of the organs in the diagrams are identical with those of the female figures. For comparison, a photograph has been made of the broadside of the female with the first or external flap lifted up, in order to show the layers or flaps representing the internal organs.

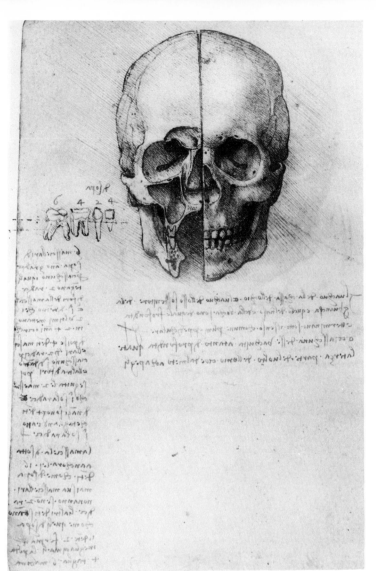

276. Two halves of a skull as drawn by Leonardo.
Leonardo da Vinci made many detailed studies of anatomy, evidently intended for a general treatise on the frame of man. He wrote: "I have dissected more than ten human bodies, destroying all the various members, and removing even the very smallest particles of the flesh." Furthermore, he said, since "one single body did not suffice, . . . it was necesary to proceed by stages with as many bodies as would render my knowledge complete; and this I repeated twice over in order to discover the differences."

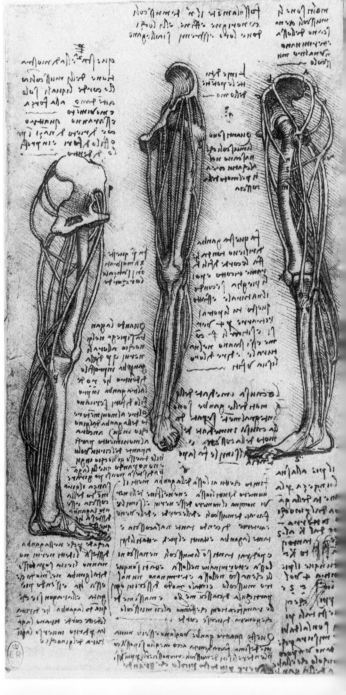

277. Leonardo's representations of muscles as threads.
In addition to his careful studies of anatomy, Leonardo devoted much thought and attention to ways of graphically demonstrating the structure of the human body. Here he uses threads to show the lines of muscular action in the leg, for he held that "every muscle uses its power along the line of its length." For clarity in complicated drawings, such as this one, he said it would be best to "first make a demonstration of the slender muscles using threads; and in this way you will be able to represent them one above another as nature has placed them; and so you will be able to name them according to the member that they serve, that is, the movement of the point of the great toe, and of its middle or first bone, and so on." Leonardo also devised ways of drawing "exploded" views of the skull and cross sections of the legs and arms.

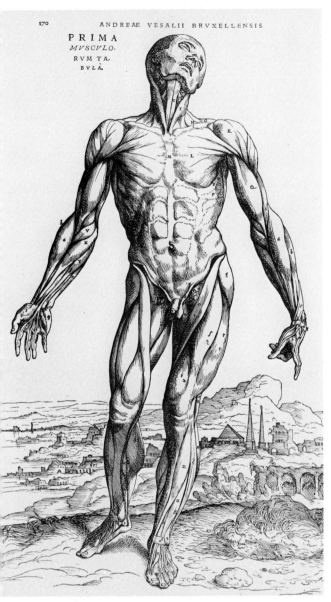

PRIMA
MVSCVLO.
RVM TA-
BVLÀ.

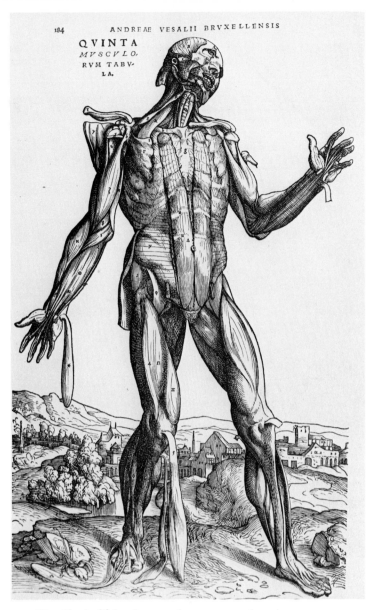

QVINTA
MVSCVLO.
RVM TABV-
LA.

278. Vesalius's first illustration of human muscles. This plate delineating the superficial arrangement of the frontal muscles was designed not only for physicians and anatomists but, as Vesalius said, to "display a total view of the scheme of muscles such as only painters and sculptors are wont to consider." If the drawings of muscles in his 1543 treatise are placed side by side in sequence, a continuous landscape is formed behind the figures of the area around Abano Terme, a region southwest of Padua.

279. Vesalius's fifth plate on human muscles. As conceived, the muscle studies done by Vesalius follow the sequence of dissection stage by stage. J. B. de C. M. Saunders and Charles D. O'Malley have pointed out that this illustration shows "the extension of the rectus abdominis muscle to the level of the first rib," although "this arrangement is only found in lower forms." While this might appear to be a "glaring blunder," it demonstrates the way in which Vesalius was apt to incorporate "many of his observations on comparative anatomy in his illustrations of the human figure." In his text, Vesalius presented "theoretical arguments as to what position a structure found in animals would occupy if it occurred in man." In this way, he made clear the difference between animal and human forms, thereby refuting earlier claims made by Galen to the contrary and calling attention to Galen's errors. Hence, as Saunders and O'Malley warn us, these illustrations cannot be fully understood without specific reference to the text. In this case Vesalius "specifically states that the upper part of the muscle is a tendinous extension found in apes and dogs," not in man.

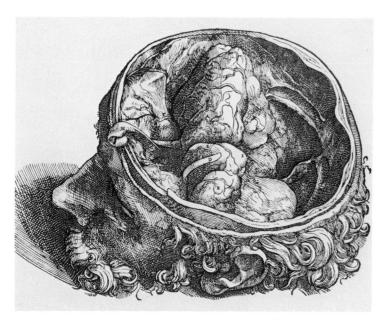

280. Man's brain. In describing his representation of the human brain, Vesalius wrote, "In this figure the head is represented from the left side, the right being somewhat elevated. Here we have evulsed the cerebellum from the cavity of the skull" so that only the lower portion of the brain is visible. Vesalius notes that "this portion of the brain does not lie in position but has been elevated . . . so that the [olfactory] processes of the brain, in form not unlike nerves, and which subserve smell, may come into view." Vesalius follows Galen in not classifying the olfactory nerves with the cranial nerves.

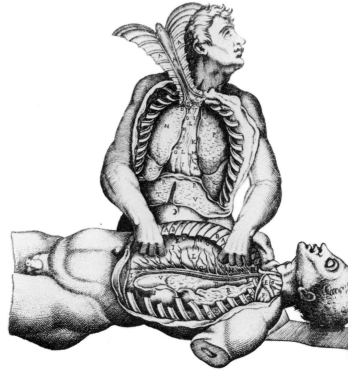

281. Man contemplates himself. This engraving of a partially anatomized cadaver anatomizing another cadaver is one of Gaspar Becerra's illustrations from Juan de Valverde's *History of the Construction of the Human Body* (1556). It was copied in many seventeenth and eighteenth century books. Although many of the plates in Valverde's book are based on illustrations in Vesalius's 1543 treatise, he introduced numerous corrections and additions. It has been said that this was the most widely read and used scientific book of the Renaissance.

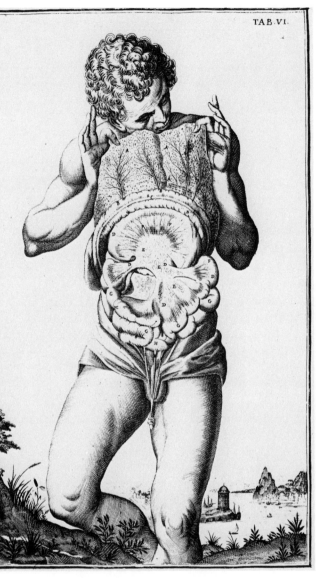

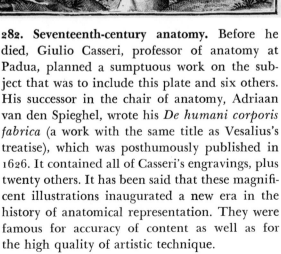

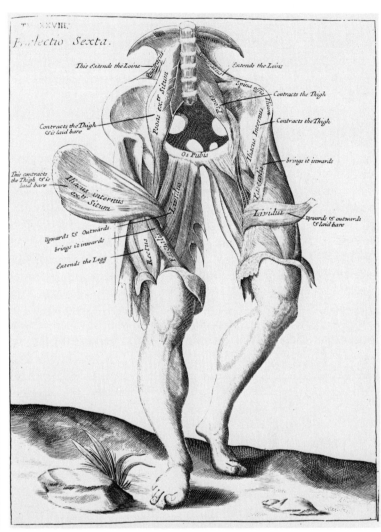

282. Seventeenth-century anatomy. Before he died, Giulio Casseri, professor of anatomy at Padua, planned a sumptuous work on the subject that was to include this plate and six others. His successor in the chair of anatomy, Adriaan van den Spieghel, wrote his *De humani corporis fabrica* (a work with the same title as Vesalius's treatise), which was posthumously published in 1626. It contained all of Casseri's engravings, plus twenty others. It has been said that these magnificent illustrations inaugurated a new era in the history of anatomical representation. They were famous for accuracy of content as well as for the high quality of artistic technique.

283. A seventeenth-century representation of muscles. Using a naturalistic background in the general manner originated by Vesalius, but without his artistry, John Browne illustrated the pelvic and upper thigh muscles of a man in *Myographia nova* (1684). Needless to say, the effects of the strolling, partly anatomized limbs is bizarre in the extreme. This plate was derived in part from the work of Casseri.

284. Using the camera obscura as an adjunct to artistic anatomy. This illustration from William Cheselden's *Osteographia* (1733) depicts the way artists used the camera obscura in drawing the human body. An artist is seated at one end of a long wooden box that has been coated on the inside with black paint. The front of the box is covered with a tightly fitted board, into which a biconvex lens is set, and the object to be drawn is placed in front of the lens and projected on a pane of glass with a matte surface. The image is focused by adjusting the distance of the object from the box. Thus, an exact replica of the image, in this case a skeleton, can be drawn by the artist on the glass and then transferred to paper or to a copperplate for engraving. Alternatively, a piece of transparent paper could be used instead of the glass surface to record the image directly. Because the lens reverses the image, the skeleton is hanging upside down, so that it will appear rightside up before the illustrator's eyes.

285. An anatomical illustration for Albrecht von Haller. One of the greatest eighteenth-century anatomy books was Albrecht von Haller's *Icones Anatomicae* (or *Anatomical Illustrations*), published in four parts in Göttingen from 1743 to 1756. Haller had studied under Bernhard Siegfried Albinus (see Ill. 287). Hence, as the historian Ludwig Choulant has remarked, "he was bound to make pictorial representations of anatomical preparations the main object of his care." One of the artists who worked closely with Haller was Joel Paul Kaltenhofer. Kaltenhofer's meticulous care in the preparation of anatomical illustrations can be seen in this first wash drawing of the "*Arteria pudenda interna*," or the pelvic artery.

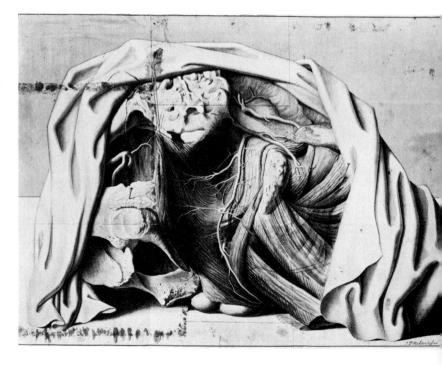

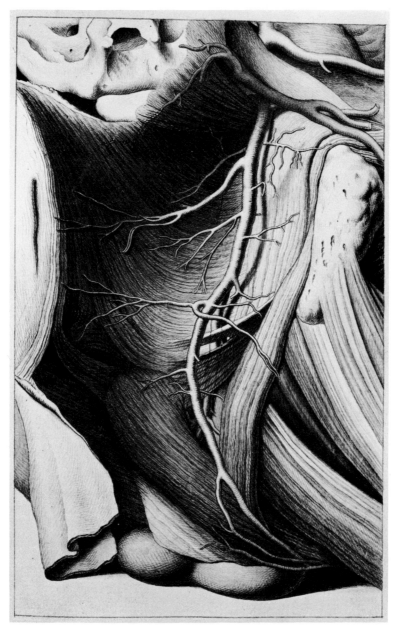

286. The final form of the illustration for Albrecht von Haller.
The first drawing by Kaltenhofer (Ill. 285) was made from Haller's
anatomical preparation and shows much more than the artery and
its immediate surroundings. In the final illustration, shown here,
the artist has concentrated on the artery itself. This served as the
basis of the copperplate engraving from which prints were made
for Haller's book. It appears as the "*Arteria pelvis*" in the fourth
fascicule of Haller's *Icones* (1749).

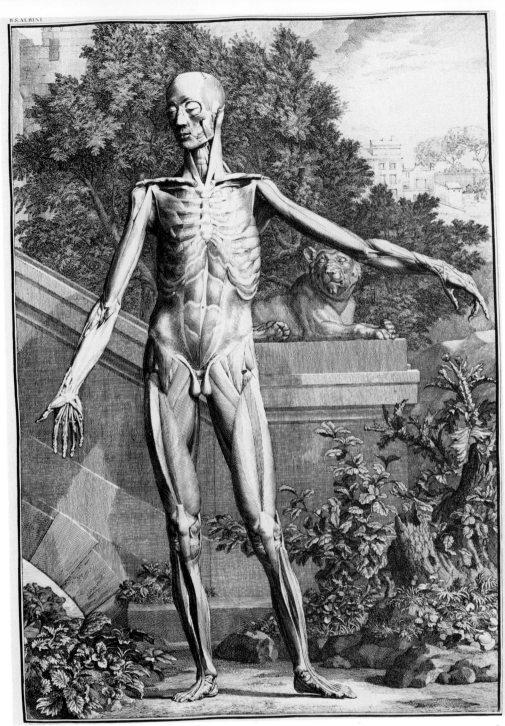

287. Pioneering anatomical studies. Albinus, a German anatomist and surgeon and professor of both disciplines at Leiden, set new standards of accuracy and thoroughness in rendering the human body. This figure is taken from his *Plates of the Skeleton and Muscles of the Human Body* (1747), and was rendered by the Dutch artist Jan Wandelaar, who did all of Albinus's illustrations. According to Albinus, Wandelaar had been "trained and guided and practically directed by me as if I myself were making the pictures and using him as a tool." No subject was illustrated that the artist "had not first thoroughly understood," and no expense was spared to achieve the best possible results. For such work Albinus set up two nets the size of the skeleton, one coarse and the other ten times finer. The coarse net was placed directly in front of the subject and the fine one was set parallel to the first and about 4 feet away. For rendering small details, the artist, who generally stood about 40 feet away, would come up close and draw what could be seen within each small square. This device, suggested by W. J. 'sGravesande, a physics professor of Leiden, was an early method for drawing perspective and kept all details in true proportion. A notable feature of these illustrations is the unchanging position of the figure, in various stages of being anatomized, against different backgrounds.

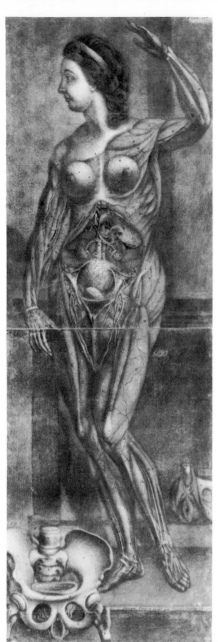

288. Anatomy printed in color. Jacques Fabien Gautier d'Agoty was a pioneer in color printing who used a variant of a printing method invented and introduced early in the eighteenth century by Jacques Christophe Le Blon. Le Blon made three copperplates, each for a different color—red, yellow, and blue—so that three separate impressions could be superimposed on one another. Le Blon produced only one anatomical plate in color, but his methods were used by Jan Ladmiral, who did a number of anatomical color prints, including some for a book by Albinus. D'Agoty's innovation was the addition of a fourth plate for black. Some of D'Agoty's prints were made to be joined together to give an anatomical representation that was in color and in life size. This figure comes from a set of eight plates issued in 1773. When pasted together, they yield four large-scale anatomical representations; this female figure, a male, a pregnant woman, and a woman in labor.

289. William Hunter lecturing on anatomy. One of the most famous surgeons and teachers in eighteenth-century England was William Hunter, who held a school for anatomy in his house on Great Windmill Street in London. Here he is shown lecturing on anatomy to artists of the Royal Academy. For his demonstration he is using a living model, a skeleton, and a statue of a "muscle man."

Pasquier in .   Joubert Sc.

ches MARC-MICHEL BOUSQUET

290. Intravascular injection. This illustration from Albrecht von Haller's *Two Memoirs on the Motions of the Blood* (1756) shows how colored fluids were injected into blood vessels so that their details could be studied anatomically. Haller was the leading physiologist and anatomist of the eighteenth century, especially famous for his concept of "irritability." This concept was used by him to describe the property of any part of the human or animal body that contracts as a result of external stimulation. Haller also made important studies of anatomy, embryology, and Swiss flora.

23

Physiology of Animals and Man

The greatest biological advance of the early Scientific Revolution was William Harvey's discovery of the circulation of the blood. He not only established a fundamental truth about the motion of blood in men and animals, but also destroyed the basis of traditional notions about physiology that went back to Aristotle and Galen. In particular, Harvey's work sounded the death knell of the doctrine of the four "humors," which was propounded by Galen and formed the basis of medical therapeutics. According to that theory the animal and human body is made up of four "humors" or fluids: blood (*sanguis*), phlegm (*pituita*), yellow bile (*chole*), and black bile (*melanchole*). These were, in turn, associated with the four Aristotelian elements (air, water, fire, and earth), and their excess or imbalance produced the four human temperaments: sanguine, phlegmatic, choleric, and melancholy. A Galenic physician would attempt to cure diseases by balancing these humors, or getting rid of an excess of one or another by bloodletting or administering a laxative or a purgative.

Harvey's work also paved the way for the rejection of Galen's notion of the three "spirits." The first of these, the "natural spirit," was allegedly added to blood in the liver, where blood was supposed to originate. From there it was carried in the veins by an ebb-and-flow movement throughout the body. The second, or "vital spirit," was said to be derived from the heart, where it combined with blood that supposedly trickled through channels or pores in the septum or muscular wall separating the right from the left ventricle.

In Galenic physiology, a difficult thing to account for was the existence of blood in the arteries. The "animal," or third spirit, was thought to be supplied to the arterial blood in the brain. Harvey showed by a quantitative demonstration that the concept of blood constantly formed in the liver was erroneous; and his doctrine of circulation precluded the possibility of blood trickling or oozing from the right ventricle of the heart to the left.

Vesalius, whose treatise on anatomy was

largely inspired by the need to correct the errors of Galen, naturally had balked at the idea of pores in the interventricular wall. In the first edition of his *On the Fabric of the Human Body* (1543), he wrote that the "septum of the ventricles is formed from the very densest substance of the heart." He observed that this wall "abounds on both sides with pits," but reported that none of these pits, "so far as the senses can perceive, penetrate from the left to the right ventricle." He therefore concluded, "We are thus forced to wonder at the art of the Creator by which the blood passes from the right to the left ventricle through pores which elude the sight."

In the second edition of his book, Vesalius added: "Not long ago I would not have dared to turn aside even a nail's breadth from the opinion of Galen, the prince of physicians. . . . But the septum of the heart is as thick, dense, and compact as the rest of the heart. I do not, therefore, know . . . in what way even the smallest particle can be transferred from the right to the left ventricle through the substance of that septum."

A second great discovery in the life sciences made during the early days of the Scientific Revolution was that of spermatozoa in the semen of male animals. The actual role of these tiny "animalcules," as they were called, was far from clear; and it was even imagined that they might carry tiny, preformed individuals. There was considerable confusion in the seventeenth and eighteenth centuries as to whether such preformed individuals actually exist, or whether animals develop from a kind of seed. Those who espoused the latter theory of development, or epigenesis, rather than preformation, were greatly handicapped by the fact that no one could find a mammalian ovum—the expected analog of the egg of a bird or of any oviparous animal. Harvey devoted many years of research vainly looking for a mammalian ovum, but that discovery was not made until the 1820s by Karl Ernst von Baer.

Seventeenth- and eighteenth-century scientists devoted much of their energy to the study of embryology. Knowledge of human embryology was a natural by-product of the work of obstetricians, who had occasion to examine fetuses resulting from miscarriages. The greatest number of embryological studies, however, were made on chicks, because it was so easy to develop fertilized eggs in an incubator and open them at regular intervals in order to examine the stages of the embryo's development. The developmental life cycles of insects and frogs were also studied, prompted by a great fascination in learning about the mysteries of life's formation on a scientific level.

During the seventeenth and eighteenth centuries, many new phenomena were also discovered by experimenters seeking physical and chemical explanations of living processes. Notable progress was made in the understanding of respiration and sensation. The rise of modern physiology—even if it has not always been one hundred percent reductionist—has always been based on the study of the physics and chemistry of life.

One theory of physiology in the seventeenth century was iatro-chemistry, or iatro-mechanics. It was based on the Greek word *iatro*, meaning physician, and explained all physiological processes in terms of chemical reactions and the interplay of mechanical or physical processes. Iatro-chemistry was the creation of Paracelsus, who believed that health depends on the proper chemical balance of bodily fluids. This point of view was in conflict with that of the previous age, in which attempts had been made to explain living forces in terms of cosmic or celestial forces, and to account for the microcosm or individual in terms of the macrocosm or universe at large. The goal now was to "reduce" living processes or the phenomena of animate nature to inanimate chemical reactions and the Cartesian principles of matter and motion plus force.

In the nineteenth century, a debate arose between those who believed that all living processes could be totally "reduced" to chemistry and physics and their opponents who held that living processes required a special "vital force" that was not operative in inorganic processes. Advocates of these two points held that living processes require a special "vital force" that is not operative in inor- or mode of reconciliation between them.

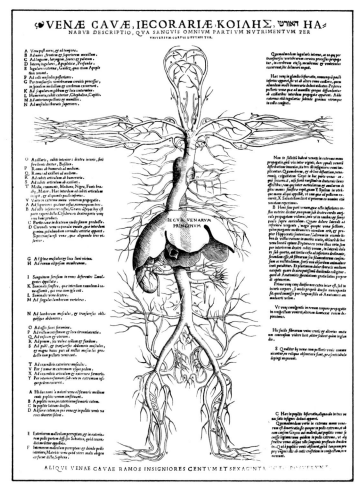

291. Vesalius's illustration of the Galenical physiology. In 1538, five years before Vesalius published his great treatise *On the Fabric of the Human Body*, a series of six plates, known as the *Tabulae anatomicae* or *Tabulae sex*, were published, intended for medical students. The first three were drawn by Vesalius himself, the second three (showing the skeleton) by the Flemish artist Jan Stefan van Kalkar, who may have been primarily responsible for the illustrations in Vesalius's great treatise. Designed for students, these plates illustrate the general state of knowledge. Thus, the liver is shown out of all proportion and with the traditional five-lobed shape; it carries a legend that the liver is the "source of the veins," in keeping with Galenical notions that blood ebbs and flows into the cava, or great vein, becoming charged with "natural spirit" in the liver. Such schemata of the venous system were of great practical importance for doctors, since bloodletting (venesection) was "the sheet anchor of therapeutics." In the original plate, both sides of the diagram are covered with remarks by Vesalius, and there is a caption reading (in Latin, Greek, and Hebrew), "A description of the Vena Cava, Jecoraria, *KOILE, HA-ORTI* or Hanabub by which the blood, nutriment of all the parts, is distributed throughout the entire body."

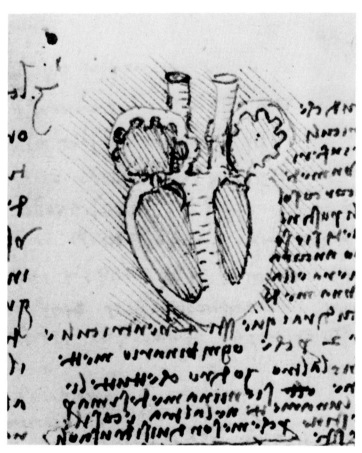

292. Leonardo's drawing of the septal pores of the heart. Leonardo da Vinci made many drawings of the heart, usually of the heart of an ox. The drawing shown here is remarkable because it shows all four cavities of the heart, the two ventricles and the two auricles. Charles Singer has noted that Galen had not paid much attention to the auricles, considering that "their only function was to prevent over-distension of the ventricles." Thus, Galen and the Galenists described the heart as if it had only two cavities, whereas Leonardo correctly shows it with four. Leonardo, according to Singer's analysis, also "succeeded in grasping quite clearly the nature and action of the valves at the root of the great arteries, and verified his views by a most remarkable experiment." He was able to prove "that the valves allowed the blood to pass in only one direction, and prevented its regurgitation." What is of greatest interest in this drawing, however, is that Leonardo actually shows the existence of pores in the septum, or interventricular wall, that separates the right ventricle from the left. It was through these pores, according to Galenical physiology, that blood could pass from the venous system into the arterial system. But there are no such pores in nature.

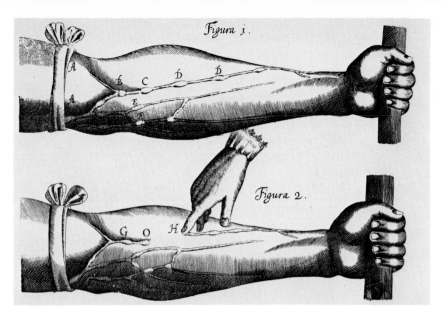

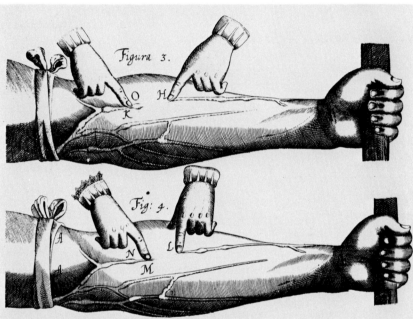

293. A demonstration of the one-way flow of blood through the veins. William Harvey's experiments shown here, which anyone can easily repeat for himself, show that blood passes through the valves of the veins only in one direction, from the extremities toward the heart. With the flow of blood temporarily stopped by a tourniquet, one can see nodes or swellings (*B,C,D,D,E,F,* in Figure 1) at each valve in the veins. An experimenter can rub his finger along a vein (as from *O* to *H* in Figure 2), "milking the vein downward"; he will observe that the valve (at *O*) prevents the blood from flowing backwards, away from the heart. "Keeping the blood thus withdrawn" (from *O* to *H*) by a finger of his right hand, the experimenter next takes a finger of his left hand and tries (Figure 3) to press blood (*K* to *O*) through the valve (toward *H*); he will find "the blood completely resistant to being forcibly driven beyond the valve" in a direction away from the heart. In a final experiment (Figure 4), two fingers, one of each hand, are placed together at a point in a vein below a valve (*N*). Keeping the finger of his left hand in place (at *L*), the finger of his right hand strokes the vein in a direction toward the valve, and toward the heart. The blood passes easily through the valve, toward the heart, leaving a stretch of vein empty, since the blood cannot get back into the vein through the valve. Thus, Harvey showed that blood passes through valves in the veins toward the heart, but not in the other direction. Hence there is no ebb and flow of blood in the veins, but only a one-way flow toward the heart. Accordingly, the blood entering the veins must come from a source outside the venous system, namely the arteries. This was only one of the many demonstrations given by Harvey in his famous book of 1628 entitled *Movement of the Heart and Blood in Animals: An Anatomical Essay,* in which he set forth the doctrine of the circulation of the blood.

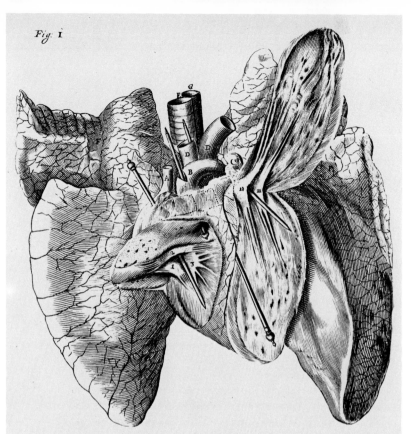

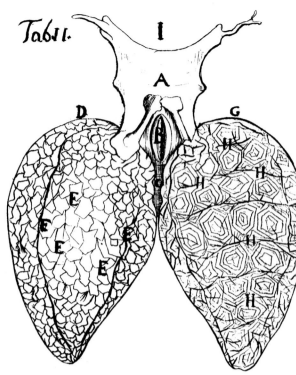

294. Anatomy of the heart. In 1629, while he was in Amsterdam, René Descartes made many anatomical studies of the organs of animals, particularly the heart. The original Latin edition of Descartes's *Treatise of Man* (1662) contained this illustration. A pair of paper flaps are pasted on the plate and folded down to show the external appearance of the heart. When the flaps are lifted, as here, the heart's internal structure is revealed. The pins are an aid to understanding the construction of the organ.

295. Discovery of the capillaries. Harvey established the fact that in human beings the blood is pumped out of the left ventricle of the heart through the aorta and then into lesser arteries, and returns to the heart through the venous system into the vena cava and into the right auricle. Having completed what is known as the major circulation, the blood then passes through a valve from the right auricle into the right ventricle, and then begins the lesser circulation, passing through the pulmonary artery into the lungs, and then from the lungs back through the pulmonary vein into the left auricle. It then moves through a valve from the left auricle into the left ventricle, and is pumped out again through the aorta to circulate through the upper and lower parts of the body. Harvey never found the actual links between the arterial and venous systems, although he did observe that the arteries and veins, by continual branching, become smaller and smaller. In 1661 Marcello Malpighi published his *Anatomical Observations on the Lungs*, in which he described his discovery of the capillaries of the lesser, or pulmonary, circulation. Thus, he was the first scientist to make known, if not the first man to actually see, how the blood passes from arteries to veins. Malpighi made his discovery in a frog's lung, an organ which is almost transparent and has "particularly conspicuous capillary vessels." This remarkable discovery was made with the aid of a microscope.

296. Two forms of circulation. The frontispiece to Sachs von Lewenheimb's book on the microcosm and macrocosm (1664) compares the recently discovered circulation of the blood in man (*below*) and the continual circulation of water between heaven and earth (*above*). Water rises up as vapor to form clouds that, in turn, produce rain, which falls into the oceans.

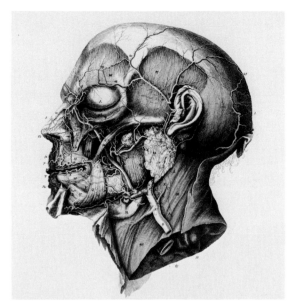

297. Circulation of the blood in the head and brain. In his studies of circulation, Haller was particularly concerned with the effects of respiration. For example, he observed the way "the blood wells up in the veins of the head, neck, chest, and abdomen" during respiration, a phenomenon most easily visible in a "dissected brain." Haller was thus able to cast out of science the earlier and incorrect ideas according to which the "movement of blood in the brain" was thought to be evidence of "contractions of the dura mater," conceived as pumping "nervous fluid through the body, in analogy to the distribution of the blood." The illustrations accompanying Haller's work show an exquisite delicacy and accuracy in the rendering of details as well as in the general structure of the circulatory system.

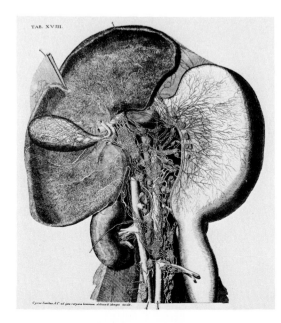

298. The lymphatic vessels. Paolo Mascagni, one of the most celebrated anatomists of the eighteenth century, was especially known for his study of the lymphatic vessels. He used injections of mercury in order to make the vessels stand out, and "establish that every vessel must in its course enter one or more lymph glands." Approximately half of the lymphatic vessels now known were discovered by Mascagni. In his study of the blood vessels and the lymphatic system, he used colored injections as well as mercury. At the height of his career, he accepted a professorship at Florence to teach anatomy, physiology, chemistry, and anatomy for artists. He also prepared a series of life-size figures to show the various systems of the human body for the training of students of sculpture and painting. The astonishingly high level of Mascagni's technique is revealed in this plate, taken from his *History and Iconography of the Lymphatic Vessels of the Human Body* (1787).

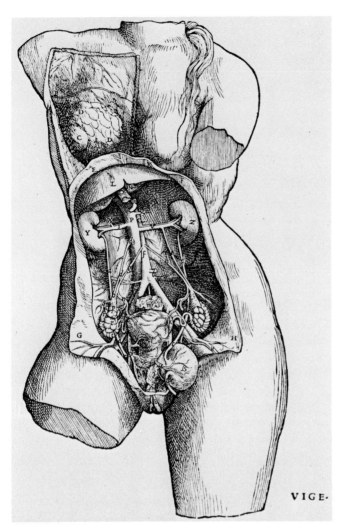

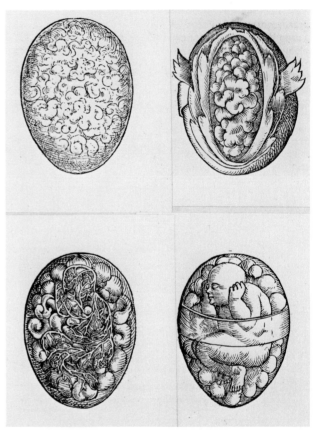

299. Generative organs of the female. On this figure, Vesalius wrote: "We have removed the skin from the right breast . . . so that the nature of the breasts might be exposed to view as far as possible. Then we have cut away the stomach, the mesentery with the intestines, and the spleen, while leaving the rectum. . . . " Furthermore, "we have stripped the external covering from the uterus . . . excising very carefully all membranes wherever possible so that the vessels carrying down seminal material to the testes [i.e., the ovaries], and others, . . . might come into view." Among other preparations, "we have excised in this figure a portion of the pubic bone so that the neck of the uterus [i.e., the vagina] and the neck of the bladder [urethra] might be seen more conveniently." As a realistic touch, a lock of hair hangs over the left breast, even though the figure is decapitated. A comparable dissected figure showed the male organs of generation.

300. Human embryology from a sixteenth-century point of view. Jacob Rüff's ideas of mammalian and human embryology followed those of Galen and Aristotle, and were published in Latin in *The Conception and Generation of Man* (1554). The English translation, entitled *The Expert Midwife*, was widely used and went through many editions. Here four different stages of the development of a fetus have been selected from Rüff's book to show the general progression. The figure at upper left represents a mixture of semen and menstrual blood in the womb. At upper right, the same mixture appears in the uterus, "wrapped round with three coats, amnion, chorion, and allantois," according to the eminent scientist and historian Joseph Needham, who also calls it "a lamentable but interesting misrepresentation of the facts." Development then proceeds as the heart starts to send out blood vessels that spread out over the embryo. Needham concluded that Rüff "must either have opened hen's eggs himself and seen the early growth of the blastoderm or have been told about it by some observer." The last two stages are even more imaginative. Rüff was more successful in his "excellent diagrams of the foetus *in utero*," showing positions "familiar to obstetricians," for whom his book was written.

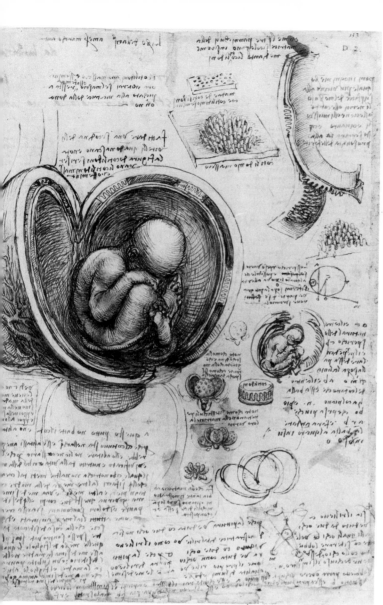

301. A developed human fetus. This drawing by Leonardo of a human fetus in the uterus is part of a page containing other drawings, including another small sketch of a fetus.

302. A human infant at birth, shown with placenta and membranes. This illustration was prepared by Giulio Casseri and was later published in Adriaan van den Spieghel's tract *On the Formation of the Fetus* (1626). "The human placenta," as George Corner observes, "is an object of considerable size, as shown in this fine old engraving from Casseri."

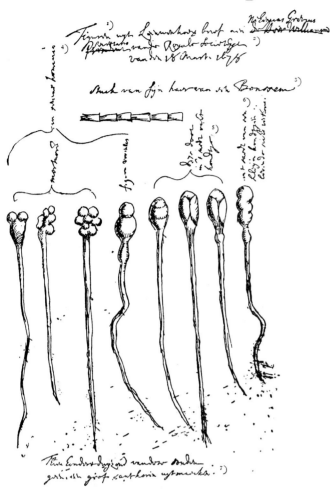

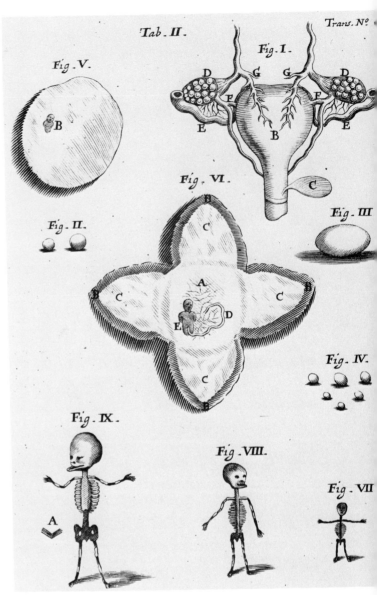

303. Spermatozoa seen through the microscope. This series of pen-and-ink drawings was made in 1678 by Christiaan Huygens, the famous physicist, after drawings made by Antoni van Leeuwenhoek. At the top of the page Huygens has written: "Figures copied from the letter of 18 March 1678 from Leeuwenhoek to Nehemiah Grew, Secretary of the Royal Society." Above the horizontal figure, Huygens has written "Morsel of hair of a polecat." Of the spermatozoa shown here, only the fourth from the left and the one at the extreme right, both with curvy tails, were apparently drawn from living material. Four are spermatozoa of the rabbit and four are of the dog. Above the figure on the extreme right, Huygens has written "From the sperm of a small dog, seen to be alive." At the very bottom of the page, Huygens has written that "ten hundred thousands of these objects will not make up a large grain of sand." Although Leeuwenhoek was the discoverer of spermatozoa, it was Huygens who was the first to publish an account of this discovery, which he declares "*was made in Holland* for the first time"; this discovery, he says, "seems very important, and should give employment to those interested in the generation of animals."

304. Preformation in man. In 1671 T. Kerckring published a figure of a preformed individual in an ovum that was three, or at most four, days old. The head and body were clearly differentiated, but there was no sign of limbs. A fortnight after conception, in the illustration shown here, the form of a homunculus, or preformed individual, can plainly be discerned, with the umbilical cord, the head and face, and the limbs all visible. Indeed, Kerckring held that this fifteen-day-old human fetus had discernible eyes, nose, mouth, and ears. "Who would have believed," he wrote, "had the knife of the anatomist not disclosed it, that the cradle of man no less than of birds was to be found in the egg." Kerckring also showed the skeletons of infants, allegedly three, four, and six weeks after conception. Although Kerckring's imaginative results received considerable circulation and were even published in the *Philosophical Transactions of the Royal Society* in 1672, they were not generally believed.

305. Paracelsian "anatomical furnace." Paracelsian medicine, following the teachings of Paracelsus, opposed the traditional medieval practice of diagnosing a patient's disease merely by the color of his urine. Paracelsus held that analysis of the urine could be useful in diagnosis only if there were also some chemical tests. In a Paracelsian treatise of 1577, the author explains (in a paraphrase by Walter Pagel) that "it is necessary to subject urine to distillation in a carefully gauged measuring cylinder, the parts of which correspond in length and width to those of the human body." The cylinder corresponding to the human body is shown in this illustration "of the anatomical furnace," which provides a means whereby the body might be "chemically dissected" for the benefit of the patient. This is the sense of the full title of the treatise, *Anatomy, that is, the Dissection of the Living Body: Or of Distillation of the Urine.*

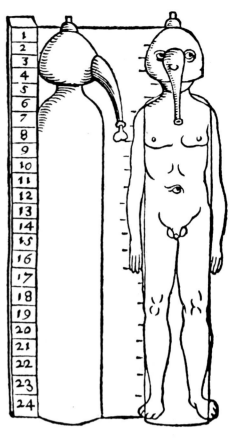

306. Organs of the human body compared to parts of a house. This comparison of chemical and mechanical processes with human physiological functions typifies the ideals of the iatro-chemical and iatro-mechanical school. In a book in Hebrew published in Venice in 1708, iatro-mechanist Dr. Tobias Cohn compared the eyes to windows at the top of a house, the shoulders to the roof, the stomach to the kitchen, and the kidneys to the water system of the house.

Et pource que l'Acquife eft dependante de toutes les au- naturelle
tres circonftances qui changent le cours des Efprits, ie ou acquise
la pourray tantoft mieux expliquer. Mais afin que ie filets qui
vous die en quoy confifte la Naturelle ; fçachez que la fubftan-
Dieu a tellement difpofé ces petits filets en les formant, ueau.
que les paffages qu'il a laiffez parmy eux, peuuent con-
duire les Efprits, qui font meus par quelque action par-

307. Mechanical response to the heat of the fire. In his *Treatise of Man*, Descartes illustrates the way heat causes motion of "spirits" in the nerve tubes, producing a mechanical action on the tubes in the brain and, in turn, affecting the pineal gland. The pineal gland then puts into motion the fluid that makes the muscles draw the hand away from the fire. Descartes extended his mechanical explanations to voluntary and involuntary muscular actions, the cause of sleep, and to automatic "reflexes."

308. The chemical basis of respiration. This is the title page of Jan Swammerdam's *Physico-Anatomical Medical Treatise on Respiration and the Functions of the Lungs* (1667). A landmark in the history of physiology, the book was Swammerdam's "inaugural dissertation." In many experiments, he studied the role of air in respiration. He demonstrated the mechanism of mammalian respiration by attaching an air bellows to the windpipe and lungs and observing the effects of inflating and deflating the lungs when the bellows were successively compressed and expanded.

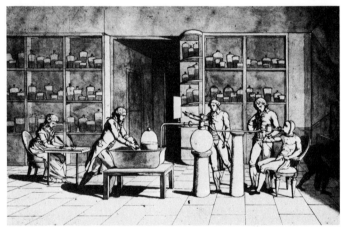

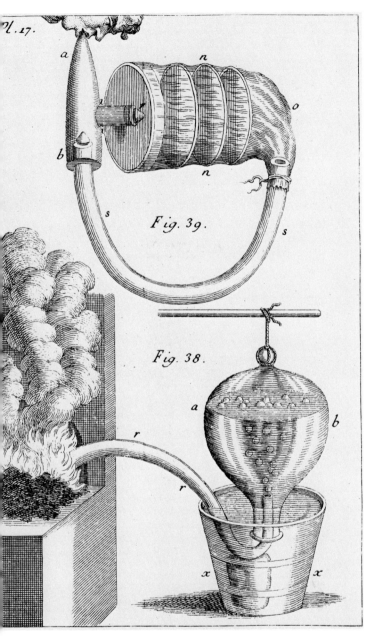

309. Hales's "rebreathing" experiment. The upper part of this plate illustrates the "Analysis of the Air," in Stephen Hales's *Vegetable Staticks* (1727). In the experiment a man breathed and rebreathed the air in a syphon. Hales studied the nature of this rebreathed air and attempted to determine how it had been changed. He wrote of different kinds of air (or "airs"), not recognizing that air is a mixture and that respiration produces a chemical change of one of its constituents, oxygen. Hales's pneumatic trough, a device for collecting gases by the displacement of water, is seen in the lower part of the plate. Gas, which has evolved from a retort heated in a furnace, bubbles up through the water into a large, inverted flask submerged in a bucket of water.

310. Lavoisier's experiments on respiration. The great chemist Antoine-Laurent Lavoisier, often called the father of modern chemistry, also investigated the chemistry of respiration. In this pair of drawings by his wife, Lavoisier is standing while she sits at a desk, pen in hand, keeping track of an experiment. Lavoisier's friend Armand Séguin is wearing a mask, which covers his face and nose and is connected to chemical apparatus. According to Lavoisier's theory, as Séguin recycled air, he used up the oxygen it contained and converted the oxygen into carbon dioxide. These illustrations were made from photographs of the original drawings, which disappeared after World War II.

227

MAMMALIA.

ORDER I. PRIMATES.

Fore-teeth cutting; upper 4, parallel; teats 2 pectoral.

1. HOMO.

Sapiens. Diurnal; varying by education and situation.

2. Four-footed, mute, hairy.   *Wild Man.*

3. Copper-coloured, choleric, erect.   *American.*
   Hair black, straight, thick; *nostrils* wide, *face* harsh; *beard* scanty; *obstinate*, content free. *Paints* himself with fine red lines. *Regulated* by customs.

4. Fair, sanguine, brawny.   *European.*
   Hair yellow, brown, flowing; *eyes* blue; *gentle*, acute, inventive. *Covered* with close vestments. *Governed* by laws.

5. Sooty, melancholy, rigid.   *Asiatic.*
   Hair black; *eyes* dark; *severe*, haughty, covetous. *Covered* with loose garments. *Governed* by opinions.

6. Black; phlegmatic, relaxed.   *African.*
   Hair black, frizzled; *skin* silky; *nose* flat; *lips* tumid; *crafty*, indolent, negligent. *Anoints* himself with grease. *Governed* by caprice.

Monstrosus Varying by climate or art.

1. Small, active, timid.   *Mountaineer.*
2. Large, indolent.   *Patagonian.*
3. Less fertile.   *Hottentot.*
4. Beardless.   *American.*
5. Head conic.   *Chinese.*
6. Head flattened.   *Canadian.*

The anatomical, physiological, natural, moral, civil and social histories of man, are best described by their respective writers.

Vol. I.—C   2. SIMIA.

311. Man in the scale of nature. This page comes from the English edition of Carolus Linnaeus's *Systema naturae* (1735), published in an English translation as *A General System of Nature* (1800). One of the radical innovations made by Linnaeus was to classify man as an animal, *homo sapiens*, characterized by his thinking and sapience (or wisdom). The different varieties of "man" were identified by external physiognomic characteristics and by habits and actions. Linnaeus, an ethnocentric European, believed that Africans were crafty and indolent, Asiatics, haughty and covetous; Europeans, on the other hand, were acute, inventive, and gentle. Linnaeus placed the apes in the same order as man and admitted that he did not know a single "generic character . . . by which to distinguish between Man and Ape." He did not know how properly to classify chimpanzees and the "tailed men" reported by explorers. In the twelfth edition of *Systema naturae*, Linnaeus concluded, "It is remarkable that the stupidest ape differs so little from the wisest man, that the surveyor of nature has yet to be found who can draw the line between them."

24

Man, a Primate

In the sixteenth and seventeenth centuries the contact made by Europeans with different sorts of men and women all over the world gave rise to a great interest in the nature and customs of non-Europeans. By the eighteenth century, a considerable amount of knowledge had been accumulated about the characteristics of natives of North America, South America, Africa, and Asia, but there was still confusion about certain reported man-like creatures who were allegedly actually types of human beings. This became a major issue when Linnaeus, in the early eighteenth century, classified man as one of the primates, notably placing him in the same "order" as apes and chimpanzees.

As black slavery became widespread, there were attempts to justify this practice politically and socially on the grounds that black people are "inferior" to people with lighter skins. Attempts were made to associate this "inferiority" with physical characteristics, such as skull shape. At the same time, a number of the true founders of physical anthropology, such as Peter Camper and Johann Friedrich Blumenbach, were violently opposed to this form of racism, which they considered to be an abuse of the rising science of man.

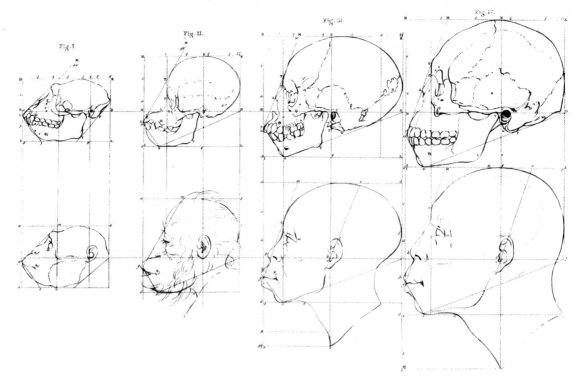

312. The facial traits of men. The eighteenth-century Dutch anatomist Peter Camper believed it especially important to consider the "facial line" and "facial angle" when studying skulls. The facial line, as seen in these drawings from his book *Physical Dissertation on the Real Differences Presented by the Facial Traits of Men of Different Countries and Different Ages* (1791), is an imaginary line drawn from the forehead to the upper lip. Another line drawn from the "auditory opening" of the ear to the base of the nose forms an angle with the facial line known as the facial angle. Here Camper compares the facial line and facial angle of a monkey with those of an orangutan, a Negro, and a Kalmuck (Buddhist Mongol). Some writers used Camper's idea of a gradation of head forms among animals to support the thesis that there were several human species, the Negro being the lowest in the scale of beings, though Camper himself did not believe this.

313. Gradation in man and in various animals. Influenced by Camper's book, Charles White produced *An Account of the Regular Gradation in Man* in 1799. In it he made this comparison between the characteristics of the skull of a Negro, an American savage, an Asiatic, two Europeans, a head as delineated by Roman painters, and another denoted as "Grecian antique." A parallel set of figures includes a snipe, crocodile, greyhound, great southern hound, bulldog, monkey, a "Man of the Woods" (chimpanzee), and an orangutan.

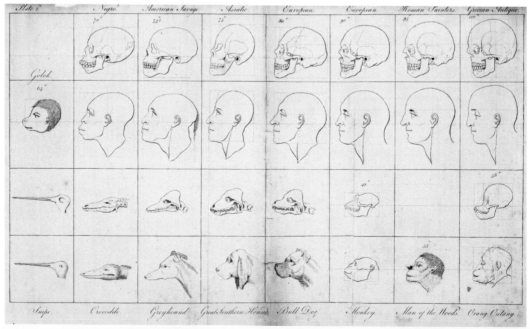

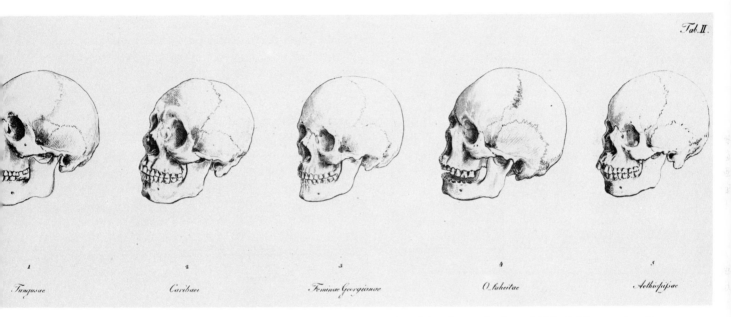

Tab.II.

1 Tungusae   *2* Caribaei   *3* Feminae Georgianae   *4* O. tahitae   *5* Aethiopissae

314. The natural varieties of mankind. In 1776 Johann Friedrich Blumenbach published his treatise *On the Natural Varieties of Mankind,* one of the founding works on the subject of anthropology. Blumenbach divided mankind into five races: Caucasian, Mongolian, Ethiopian, American (Indian), and Malayan, the basis of all later classifications. He also insisted that all the races were merely varieties of a single species. He could not accept the Linnaean concept that man and apes form a single order (which Linnaeus called the Primates), and divided the Linnaean order into two: Bimana for humans, Quadrumana for apes. The five skulls shown here represent, from left to right, a Mongolian, an American Indian, a Caucasian (a Georgian woman), a Tahitian, and an Ethiopian. Blumenbach later reproduced portraits of representative "specimens" of these five "races."

NEGRES

315. An eighteenth-century black family. In the eighteenth century there was considerable interest in the manners, customs, and physical, or physiological, characteristics of men and women from different parts of the world. This accumulated knowledge went into the foundation of comparative human physiognomy and comparative racial physical anthropology. It was obvious that men and women of different colors and "races" were similar in many ways, though their customs and manner of living differed. The disparities became important for other than scientific reasons since Caucasian Europeans made themselves masters of men and women whose skin was another color. Until well into the nineteenth century, the enslavement of black people was justified on the grounds of racial differences and the alleged racial superiority of white Europeans.

316. The passions. Buffon's *Histoire naturelle* (1749) introduces the "Natural History of Man" by describing the "Nature of Man." According to Buffon, "Though man be much interested in obtaining a knowledge of himself, yet I suspect that he is better acquainted with every other object." Buffon describes carefully the anatomy and physiology of human beings and then goes on to discuss many aspects of observable life as part of natural history. These drawings represent five different passions, as expressed in the face of a woman. Beginning with Figure 1 (*upper right*), they are sadness, terror, scorn, envy, and happiness, the latter expressed by a rather restrained smile. Figure 4 (*lower right*) does not appear to exhibit "envy" by present standards.

Part Seven

SCIENCE
AND
SOCIETY

317. Louis XIV visiting the Académie des Sciences in 1671. Sebastien Le Clerc engraved this scene of Louis XIV (wearing a plumed hat) with Jean Baptiste Colbert at his left. Colbert, the comptroller general of finances, was primarily responsible for the organization in 1666 of the French Academy of Sciences, which was partly modeled after the Royal Society. Standing behind Louis and Colbert is Claude Perrault, an architect, anatomist, physician, and former soldier. Through the open window, work can be seen in progress on the new Royal Observatory, which Perrault designed. The instruments and specimens around the room reflect the varied interests of the academy in physics, chemistry, astronomy, anatomy, natural history, and cartography.

25

The Institutionalization of Science: Scientific Academies and Societies, Museums, and Gardens

A feature of science in the seventeenth century was that men with similar interests in the physical and natural sciences began to band together in societies dedicated to the advancement of the new knowledge. Many of these early groups were short-lived, not surviving the death of a principal patron. By the 1660s, however, permanent scientific societies had been founded. The first of these, dating from 1660, was the Royal Society of London, whose full title contained the phrase "for promoting Natural Knowledge." The fellows, as the members were called, agreed to meet regularly to hear reports of scientific work done in England and other parts of the world and to see experiments being performed. The fellows decided that no discussions of politics or theology would be allowed because these subjects were divisive.

The Royal Society hired Robert Hooke to perform weekly demonstrations of new experiments. Henry Oldenburg, the first secretary, wrote letters about the activities of the society to correspondents all over the world; he found the writing of these epistolary reports so time-consuming that he began to publish a journal, *Philosophical Transactions*, which has been published continuously for over three hundred years.

Two months before the first issue of *Philosophical Transactions*, the inaugural issue of the *Journal des Sçavans* ("Journal of the Learned") was published in France. This journal was associated with the Paris Royal Academy of Sciences, which had been founded in imitation of the Royal Society. A major difference between the two academies was that Jean Baptiste Colbert, the chief finance minister of Louis XIV, provided government funds for the support of the activities of the Paris Academy of Sciences, whereas the Royal Society did not receive any money from the government. The Royal Society financed its activities through fellows' dues and contributions. In Paris, financial support was available for the library, for the performance of experiments, and for the payment of salaries to the principal academicians. Again following the example of England, where the Royal Greenwich Observatory had been established in 1675, the Paris Observatory was also founded under royal patronage.

By the eighteenth century, there were similar scientific societies throughout Europe—Berlin, Bavaria, Denmark, various parts of Italy, and Russia. America's oldest scientific or learned society, the American Philosophical Society, Held in Philadelphia for the Improvement of Natural Knowledge, dates from the 1740s. Some of these societies, such as those in Berlin and Philadelphia, have traditionally included among their members scholars devoted to the humanities and social sciences as well as those working in the natural and exact sciences.

Through their meetings and journals, the scientific and learned societies helped disseminate new knowledge and published reports of experiments, observations of plants and animals or of unusual natural phenomena, and book reviews. Election to such a society or academy was a formal mark of distinction and often a token of the international fellowship of science, since there was usually provision for a few foreign members. The academies and societies also awarded prizes and medals for distinguished contributions, and even sponsored contests for solutions to particular outstanding problems. Among the latter were the problem of impact in the seventeenth century and, in the eighteenth century, problems concerning the tides and the nature of light.

The creation of the modern museum was a characteristic feature of the Renaissance. Museums were established for collections of art works, antiquities, and specimens of natural history, and all three subjects were often combined in the same museum in an indiscriminate manner.

It is clear from illustrations of early natural history museums that exhibits were not always designed primarily for scientific purposes.

Shells, for example, were sometimes arranged aesthetically in geometric patterns to display their intriguing shapes and beautiful colors. In the seventeenth and eighteenth centuries such displays often took the shapes of animals as well as humans. An extreme example of this trend was the naturalistic way in which the pictorial museum of Frederik Ruysch in Amsterdam arranged its human skeletons. They were placed in lifelike settings, in attitudes of both living and dying.

In the seventeenth and especially eighteenth centuries, botanical gardens were established throughout Europe, primarily for the purpose of growing medicinal plants, since at that time many drugs were derived from plant material. The gardens were often arranged for the pleasure of visitors, the most notable being the Royal Garden in Paris. In such gardens, shrubs, flowers, and trees from all over the world were often arrayed according to their country of origin.

A growing interest in natural history is especially discernible in the eighteenth century. There arose a great vogue for "botanizing" and for collecting and preserving plants, animals, and insects. Extensive collections were made of minerals with interesting shapes and colors and of precious and semiprecious stones. Cabinets of curiosities were not the exclusive prerogative of royalty and the nobility, but became increasingly common in the homes of the upper middle classes.

The general interest in natural history may be seen in the magnificent publication of the forty volumes of Buffon's *Histoire naturelle*, published in a sumptuous manner by the French Royal Press, with magnificent illustrations. The universal appeal of this work led to its being translated into English. Natural history may be considered the characteristic science of the eighteenth century, just as dynamics and celestial mechanics had been characteristic of the seventeenth century.

318. The Academy of the Lynxes. One of the first academies of science was founded in the opening year of the seventeenth century by Duke Federico Cesi of Rome, an amateur scientist who was interested in bees and plants. The lynx for which the academy (also referred to as the Lincean Academy) was named supposedly had extraordinarily sharp eyes and thus symbolized the ability of the new science to see more deeply into the secrets of nature. In 1610 the academy was reorganized on a larger scale, with plans to establish "scientific, non-monastic monasteries" all over the globe. Each monastery was to have a central museum, library, printing office, botanical garden, and laboratories. One of the opening pages of the academy's record book portrays its symbol, the lynx, surmounted by the crown of the Cesi family. One of the members was Galileo, who proudly announced his membership in the academy on the title page of many of his works, describing himself as "Galileo Galilei Linceo" ("Galileo Galilei Lincean Academician"). In his dialogues he always referred to himself as "our Academician."

319. The Academy of Experiment. The Accademia del Cimento (or Academy of Experiment) was founded after Galileo's death (1642) by his pupils or by pupils of his disciples Vincenzo Viviani and Evangelista Torricelli. The patrons were two Medici brothers, Grand Duke Ferdinand II and Leopold, who were amateur physicists and experimenters. They and the members of their circle were also interested in natural history; and Leopold, in particular, studied the artificial incubation of eggs. The published account of the activities of the academy stressed experiments on magnetism, the vacuum, and temperature measurement. In this imaginary reconstruction of an academy meeting, in a 1773 engraving by G. Vascellini, some of the instruments used in these experiments are shown at lower left.

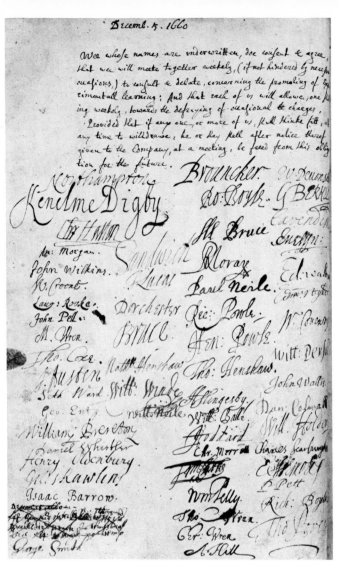

320. The founding of the Royal Society. This document, dated 5 December 1660, states that those who have subscribed their names "doe consent & agree, that wee will meete together weekely . . . to consult & debate, concerning the promoting of Experimentall learning: And that each of us will allowe, one shilling weekely, towards the defraying of occasional charges." Lord Brouncker, the society's first president, has signed his name at the top of the center column. Beneath it is Robert Boyle's signature, and second from the bottom in the center column is the name of Christopher Wren, architect and geometer. King Charles II was the patron of the society. A certificate of candidature had to be signed by eight fellows and posted before membership in the society was voted. Benjamin Franklin was made a fellow in absentia in 1753. Among its other activities, the Royal Society also published or sponsored important papers and books, including Newton's *Principia* in 1687.

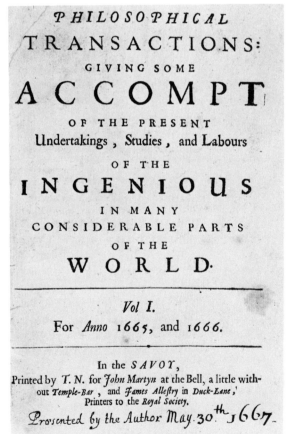

PHILOSOPHICAL
TRANSACTIONS:
GIVING SOME
ACCOMPT
OF THE PRESENT
Undertakings , Studies , and Labours
OF THE
INGENIOUS
IN MANY
CONSIDERABLE PARTS
OF THE
WORLD.

Vol I.
For *Anno* 1665, and 1666.

In the *SAVOY*,
Printed by *T. N.* for *John Martyn* at the Bell, a little without *Temple-Bar* , and *James Allestry* in *Duck-Lane*,
Printers to the *Royal Society.*

Presented by the Author May. 30th 1667.

321. Philosophical Transactions. This is the title page of the first volume of the *Philosophical Transactions*, the journal of the Royal Society of London. The first editor (and "author") was Henry Oldenburg, the first secretary of the Royal Society. The *Philosophical Transactions* is one of the two oldest learned journals in existence and has been published continuously since the latter part of the seventeenth century; the other is the French *Journal des Sçavans*.

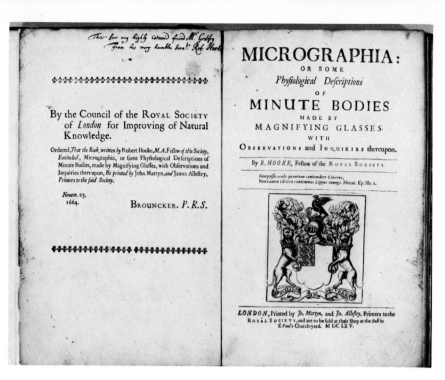

322. Hooke's *Micrographia*. The title page and autographed facing page of Robert Hooke's illustrated account of his observations made with the microscope (1665) show that the Royal Society had the privilege of "licensing" books. Lord Brouncker, P.R.S. (President of the Royal Society) signed the order of the Council of the Royal Society, dated 23 November 1664, *"That the Book written by* Robert Hooke ... *Be printed."* This illustration is reproduced from a presentation copy autographed by the author for his friend Mr. Godfry.

323. The Copley Medal. Among the activities of the Royal Society was the encouragement of science and the recognition of outstanding scientific activity, for example, by the award of a gold medal endowed by Sir Godfrey Copley. In 1753 the Royal Society awarded this medal to Benjamin Franklin. This picture of the medal showing the obverse and reverse sides, and a summary of the testimonial to Franklin on the occasion of the award were published in the *Gentleman's Magazine* in December 1753.

M. T. A. Lavoisier, Member of the Royal Academy of Sciences at Paris, and well known to the learned world for his many important discoveries in Chemistry, being desirous of admission into the Royal Society on the foreign List; we whose names are subscribed do, from our knowledge of himself or his Works, recommend him as a Gentleman whose distinguished character in Science renders him worthy of the honour he solicits.

Read Nov 8th 1787

Ballotted for & elected April 3 1788

H. Cavendish
Blagden
C. Greville
Saml. Court Simmons
C. W. Gray
Brown
Wm. Hausken
J. A. DeLuc

324. Certificate of candidature for fellowship in the Royal Society. In order to become a fellow of the Royal Society, a certificate of candidature signed by eight fellows of the society had to be posted. Lavoisier's certificate declares that he is "well known to the learned world for his many important discoveries in chemistry," and adds that he is "desirous of admission into the Royal Society on the foreign List." The signers declare that "from our knowledge of himself or his Works" they "recommend him as a Gentleman whose distinguished character in Science renders him worthy of the honour he solicits." The first signer and principal sponsor is the British chemist Henry Cavendish. The application was first read on 8 November 1787. It is stated that Lavoisier was "Ballotted for & elected April 3ᵈ 1788."

325. Members of the Académie des Sciences at work. Two views from an almanac of 1676 show members of the Académie des Sciences in Paris studying in the Royal Library (*top*) and working in the "Laboratoire" or the "Salle des Expériences" (Hall of Experiments) (*bottom*).

326. The Académie des Sciences meeting in the Royal Library. This 1686 engraving shows the members of the academy using the books and globes of the king's library. In the background is an imposing throne on a dais.

327. Blank certificate of the French Academies. Louis XIV established two royal academies, the Academy of Sciences and the Academy of Inscriptions and Fine Arts. In this engraving the dominant figure, seen at the top, is Louis himself, the "Sun King," with the sun over his head, a lyre at his right and a caduceus at his left. In the lower central portion, the figure at right holds an open book, with the titles *Journal des Sçavans* and *Mémoires de Physique.* Medallions along the right-hand side show the scientific and learned societies and institutions: the Royal Observatory, the French Academy, and the Royal Academy of Sciences. At the bottom left, there is a portrayal of a school of navigation; at the bottom right, the academy of music.

328. The Académie des Sciences in the seventeenth century. This extremely rare view of the Paris Academy of Sciences shows how the scientists gathered together to discuss the questions of the day.

329. A South American view of the activities of the Académie des Sciences. This certificate was prepared to celebrate the doctorate of a South American Jesuit, Carolus (Charles) Arboleda. Charles Marie de La Condamine records in his account of his voyage south of the equator (Paris, 1751) that while he was at Quito, he was invited to attend the defense of a thesis in theology that had been dedicated to the Paris Academy of Sciences. La Condamine was given a print made from an engraving on a silver plate. He presented the engraving and a Latin poem to the academy on his return and had a copy made of the engraving to illustrate his book of travels. The central dominating figure is Minerva, and the various children were intended to show the activities of each of the classes of membership in the academy. These include natural history, microscopy, the study of fire and flame, chemistry, mathematics, geography, surveying, and astronomy. At upper left can be seen the Paris Observatory and a large telescope.

ILLIVS
SVB VMBRA.

MOESTI COMITAMVR HONORE.

Kisbenhavn, trykt udi det Kongelige Waysenhuus, 1752.

330. Meeting of the Royal Danish Academy of Sciences. This eighteenth-century print illustrates a meeting of the Danish Academy on 18 January 1752. Although a relatively small country, Denmark has always produced a surprisingly large number of the world's greatest scientists. The Latin expression "Moesti comitamur Honore" comes from Virgil's *Aenied,* from Aeneas's speech at the bier of Pallas, and indicates that the society had gathered in sorrow to do honor to someone who had died. The phrase at the top means "In his shadow."

331. The Imperial Academy of Sciences at St. Petersburg. Peter the Great founded the Russian Academy of Sciences in 1724. This late eighteenth-century illustration shows a small pavilion with a cupola that housed a globe, which people could enter to study a depiction of the heavens. It was located in the square behind the academy's building on Vasil'evskii Island. The caravan crossing the square gives some idea of the scale and size of the globe and the pavilion that contained it.

332. The Calceolarian Museum. This museum was founded in Verona in the second half of the sixteenth century. Here, in neat array, specimens of all sorts are arranged on shelves along the wall and in drawers. Above the top shelf on each side stands a row of stuffed birds, while various stuffed animals hang from the ceiling. The museum was based on the collections made by a doctor of Verona, Benedetto Ceruto; the descriptions of the specimens were made by Andrea Chiocco, who was responsible for the publication of an account of the museum in 1622.

Within the illustration, the following labels appear:

MUSEI WORMIANI HISTORIA
LUGD. BATAVORUM
EX OFFICINA ELSEVIRIANA
Hag̔ Tyrog̔ 1655.

METAL· · · METALLICA · · · · METALLA MINERALIA

COCHLEÆ · · · TURBINATA · · · CONCHILIA · · · MARIANA

ANIMALIUM PARTES · · · CONCHILIATA · · · VARIA

LAPIDES · · · LAPIDES

SULPHURA

SALIA

LIGNA

CORTICES

HERBÆ · · RAD

333. **The Museum Wormianum.** One of the great museums of the seventeenth century was established by Ole Worm (or Wormius), a physician in Copenhagen. Worm was an avid collector, and his museum contained man-made artifacts as well as objects of natural history. Everything on display was classified according to a set of rigorous categories of Worm's own invention; the same classification was followed in the museum's catalog published in 1655. The collection has been described as "an assortment of bizarre and exotic objects, antiques, and stuffed animals." It was notable for having many tools and artifacts (left and rear walls), a kayak (hanging from the ceiling), and a remarkable collection of many prehistoric stone implements, although at the time, the latter were not attributed to man's fabrication. The implements were called *cerauniae*, a Latin word relating to thunder and lightning, indicating they were considered to be a kind of meteorite, thought to have fallen to earth when lightning flashed. Following Worm's death, the museum was taken over by King Frederick III, who opened it to the public for a small admission fee, making it one of the first public museums.

334. An ideal conception of a cabinet of natural history. In 1719 Levinus Vincent, a Dutch collector and writer, published a description of his cabinet of curiosities, which he called a "theatre of the marvels of Nature." The most impressive thing about this picture and the others in his published account is the scale and variety of the collection. In the great room shown here, attendants are displaying trays of specimens to interested spectators, who are strolling down the main aisles. On the left wall are jars containing natural curiosities, including creatures with two heads. This artist's view is probably somewhat idealized and exaggerated and would not appear to be an accurate representation of the dimensions of the cabinet or gallery drawn to scale.

335. Natural history combined with anatomy. At the end of the seventeenth century, Frederik Ruysch established an important museum of anatomy in Amsterdam, where he was a professor of anatomy. In 1710 the museum contained more than 1,300 anatomical preparations mounted in liquid. Peter the Great purchased the entire collection on his second visit to Amsterdam and moved it to St. Petersburg in 1717. Ruysch created his public displays in order to produce a picturesque effect, rather than to illustrate either a "scientific classification of animals or organs," or a "system of philosophic anatomy." The amazing illustration shown here, published in a description of Ruysch's museum in 1691, has been described by the anatomist and historian F. J. Cole as follows: "A skeleton [at the extreme right] balances an injected spermatic plexus in one hand and a coil of viscera in the other; minatory assortments of calculi of all sizes and shapes occupy the foreground; in the rear a variety of injected vessels backed by an inflated and injected tunica vaginalis combine to form a grotesque arboreal perspective; another skeleton *in extremis* is grasping a specimen of that emblem of insect mortality, the May-fly, and a third [at rear center] is performing a composition 'expressing the sorrows of mankind' on a violin symbolized by a bundle of injected arteries and a fragment of necrotic femur."

336. Natural history and Baroque art. Collections of natural history were assembled as much to satisfy man's curiosity about the exotic forms of nature and man as to provide information for science. Shells, in particular, were admired for their beauty, and were often displayed in artistic arrangements that had nothing to do with the principles of scientific classification. This illustration from a book by the Dutch naturalist and traveler Albert Seba, published in 1758, depicts conchological birds, flowers, and baskets produced for a museum exhibit. These attempts to produce art out of natural history may have provided an additional stimulus to collecting, but had nothing whatever to do with the advancement of science. Seba's seven-headed Hydra is shown in Ill. 257.

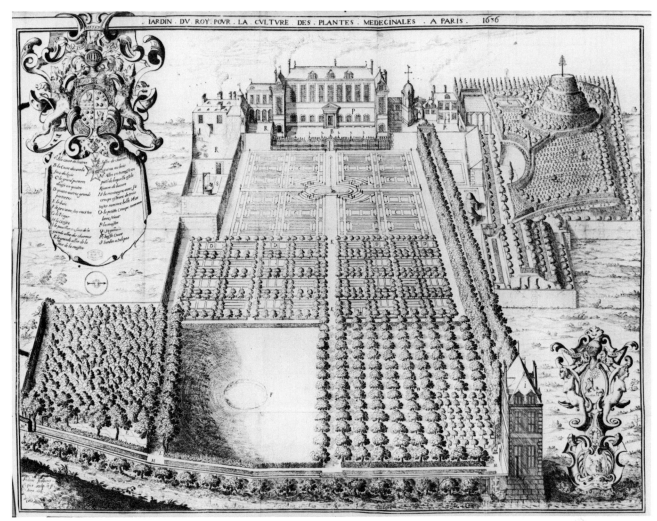

337. The Royal Garden, Paris (1636). The establishment of large-scale botanical gardens became a feature of seventeenth- and eighteenth-century science. Most famous of them all was the Jardin du Roi, founded in the 1630s and laid out in the formal geometric way made popular by Dutch horticulturists and landscape gardeners. Such gardens contained not only ornamental plants but important living botanical specimens from all over the world. The most famous superintendent of the Jardin du Roi was the eighteenth-century naturalist Buffon. After the French Revolution, the garden was renamed Jardin des Plantes, and the Muséum d'Histoire Naturelle was built along the left side of the garden. It became the world center for the study of many aspects of natural history, notably paleontology. Among the prized possessions of the museum are specimens of fossil bones and the skin and skeleton of a moose donated by Thomas Jefferson.

338. Jardin des Plantes in 1794. This lovely informal scene in the garden was captured in watercolors by Jean-Baptiste Hilair. Obviously, the public took great delight in exploring such attractive resources.

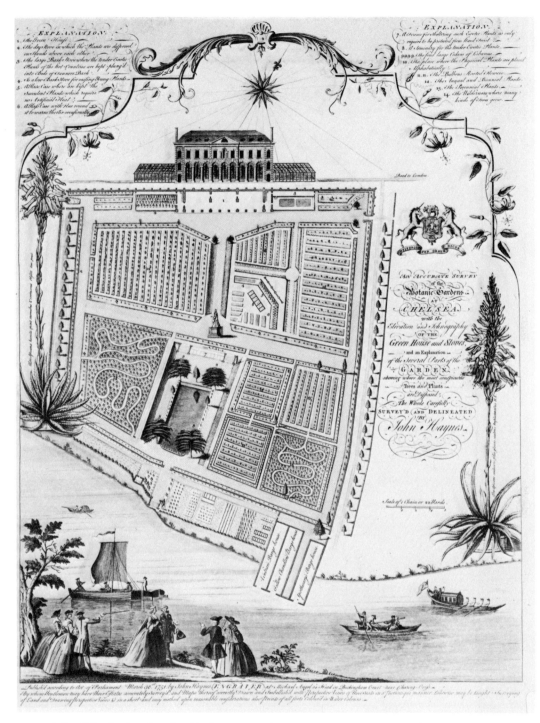

339. The Physic Garden in Chelsea. A major medical botanical garden was established in 1676 in Chelsea by the Society of Apothecaries. This plan, drawn in 1753, shows the way the gardens were laid out and the greenhouses on either side of the main building. In the description, attention was called to the special facilities "where the tender Exotic Plants of the hot Countries are kept plunged into Beds of Tanner's Bark" in a heated environment. At the lower left and lower right serpentine paths wind through "the Wilderness where many kinds of Trees grow."

HERMANNI BOERHAAVE
SERMO ACADEMICUS
DE COMPARANDO CERTO
IN PHYSICIS

LUGDUNI BATAVORUM,
Apud Petrum vander Aa, Bibliopolam.

MDCCXV.

340. Hermann Boerhaave lecturing in Leiden. Boerhaave, one of the most famous teachers and lecturers of the eighteenth century, held so many professorships at the University of Leiden that it was said he didn't occupy an academic chair but a whole settee. He was most famous for his writings on medicine and for introducing bedside teaching to accompany his formal lectures on medical care, and he also wrote a widely read and often reprinted textbook on chemistry. Here he is shown lecturing to a typically large audience in 1715.

26

Science and Society

In the early years of the new science, the condemnation of Galileo for the heresy of Copernicanism was seen as a dramatic conflict between what scientists considered to be their freedom of thought and expression and the dictates of the establishment. At that time science was thought of as directly related to man's political and religious concerns, and indeed astronomical science did serve Christian missionaries in Galileo's day in China by providing at least one area in which Western Christian Europe was obviously superior to the Far East. The scientists of the seventeenth century were convinced that the results of their studies would have important political and socioeconomic consequences by affecting communication and transportation, the conduct of military and naval affairs, mining, manufacturing, and medicine. The application of statistics in "political arithmetic" was a direct attempt to produce a science-based statecraft.

By the eighteenth century, some scientists gained prominence in the political spotlight.

Benjamin Franklin's success as the representative of the American colonies in France was due in large measure to the great reputation he had achieved as a scientist—as a leader in electrical research and as the discoverer of the principles on which was based the invention of the lightning rod. In Britain, Joseph Priestley, the noted chemist, outspokenly espoused the cause of the French Revolution. The victim of a mob attack for his political opinions, he was forced to leave the country. Across the channel, Lavoisier was brought to trial and guillotined for his role in the "tax farm," a private corporation empowered to collect taxes in prerevolutionary France. When special clemency was asked for Lavoisier, on the grounds that his scientific genius would be of use to the new republic, the presiding judge was said to have replied that the republic had no need of scientists in general, or of chemists in particular. The revolutionary government abolished the old Royal Academy of Sciences, but eventually it reorganized that scientific institution and established new schools that featured advanced science and trained teachers for the nation's new schools. Any careful study of events at the end of the eighteenth century discloses that scientists were not able to live in complete isolation from the pressures of society.

In general, the great expansion of science in the eighteenth century increased the influence of science on human affairs, and many scientists received public recognition for their discoveries. By the time of the American Revolution, Thomas Jefferson had little difficulty in gaining assent to his view that the time had come for America to take her place among the nations of the world "to which Nature and Nature's laws entitle her." He did not appeal to scripture or to divine principle, but based America's claims to justice on science and on the principles of nature revealed by science.

Scientific activities directly related to the needs of society included, as we have seen, the establishment of observatories to find a sure way of determining longitude at sea. Science also raised some practical issues for ordinary people, including the reform of the calendar to reflect the exact determination of a solar year. The introduction of the new Gregorian calendar, which was eleven days ahead of the old Julian calendar, led ignorant people in England to believe that the change in calendar had robbed them of eleven days of life. They demanded that the "lost" eleven days be restored to them.

Science was also called on to investigate new practices of questionable value that could be potentially dangerous. When mesmerism was spreading like wildfire through all classes of society, the French government asked scientists to undertake a critical evaluation of this new form of alleged therapy. Two royal commissions, one representing the Academy of Sciences and the Paris Faculty of Medicine, and the other the Academy of Medicine, investigated the mesmeric prac-

tices and cures; their negative conclusions were a major factor in discrediting the mesmerists and in causing Franz Anton Mesmer himself to leave France. The cartoons of the 1780s show how Mesmer was ridiculed, but other cartoons of the period make equal fun of such important social discoveries as Edward Jenner's method of vaccination that conferred immunity to the dread smallpox.

A notable feature of the eighteenth century was the rise of interest in science among nonspecialists. Teaching of the physical sciences in universities was improved by the introduction of new kinds of apparatus for performing lecture demonstrations, and the same sort of equipment was used by many devoted amateurs. Public lectures on science for ladies and gentlemen became popular in Europe and America.

Another development was the publication of books that explained the facts and principles of science to men and women who had no professional interest in the subject. Francesco Algarotti wrote a popular exposition specifically addressed to women that went through many editions, including translations from Italian into French and English; it was called *Newtonianism for the Ladies*. His model was Bernard de Fontenelle's very popular series of dialogues on astronomy called *Conversations on the Plurality of Worlds*.

In France the Abbé Nollet, famed for his electrical investigations, was much in demand for the performance of scientific demonstrations and experiments before the ladies and gentlemen of the court. Nollet wrote a book on the way to perform experiments and a six-volume general work on experimental physics that contains handsome illustrations, together with lucid explanations of the scientific principles such experiments were intended to display. From the illustrations it is obvious that Nollet's books were intended for the amateur. In England a series of books popularizing all aspects of the sciences was written by Benjamin Martin, who also manufactured and sold the scientific instruments he described and illustrated in these books. Some of the most beautiful scientific instruments of the eighteenth century are signed "B. Martin," and the trade in them was so lucrative that some of Martin's unscrupulous commercial rivals engraved his name on their own products in order to sell them more readily.

A popular instrument for scientific demonstration was the orrery, a mechanical model of the Copernican system, with geared mechanisms that made the planets and their moons revolve with periods having the same relation to one another as the actual periods of revolution in the heavens. No college or university, museum, or collection of instruments worthy of the name was without one, and the orrery was the prized possession of many an amateur.

The new science of electricity proved to be a particularly attractive subject for scientific lecturers, in both the Old World and New World. With explosions of houses and the production of large sparks and crackling noises, electricity lent itself to exciting demonstrations. One of Franklin's correspondents proposed to travel around the world, supporting himself by giving public demonstrations in electricity. By the end of the century, there was also a considerable public interest in chemical lectures. In London such lectures were a feature of the new Royal Institution.

The great success of Newtonian science and the elaboration of the Newtonian system of the world still inspired many people to learn more about the universe and the principles that govern it, but electricity and chemistry were new sciences for the eighteenth century, and they attracted devotees in large numbers. Very likely, natural history was even more popular than the more difficult physical sciences. This feature of the eighteenth century may be seen in the number of museums and gardens that were created and the production of expensive and lavishly illustrated books on plants that appeared. These books were obviously intended for the delectation of a wider and wealthier group than mere scientists. There was also a great vogue of botanizing, of learning to identify plants according to the Linnaean system. Private collectors were proud to display unusual minerals, living or dried plants, stuffed animals, and varieties of natural curiosities.

341. The sacking of Dr. Priestley's house. On the second anniversary of the French Revolution, 14 July 1791, a mob of British ultraconservatives burned Priestley's chapel in Birmingham, and then proceeded to destroy his home. The rioters smashed his laboratory apparatus, broke his furniture, and threw his books and manuscripts out of the windows. It is said that an attempt was made "to start a fire by means of sparks from the doctor's electrical apparatus, but without success." A dissenting minister and Unitarian, Priestley had been accused by his enemies of having "explained away" the Bible. Three years later, his overt sympathies for the French Revolution made his situation so intolerable that he emigrated to America. This lithograph, made by Charles Joseph Hullmandel, was based on a painting by a pupil of William Hogarth who had executed an on-the-spot drawing of the mob in action.

342. Lavoisier's arrest. Arrested because of his association with the notorious "tax farm," a private tax-collecting agency set up before the Revolution, Lavoisier was publicly guillotined, along with other "tax-farmers," on 8 May 1794. In 1796 ("L'An 7 de la Répub.") a counterrevolutionary broadside was published in protest against the sentence of execution, portraying this scene of Lavoisier's arrest. Lavoisier is sitting in front of his chemical apparatus, while police agents ransack his possessions, looking for incriminating documents. The broadside also contains a portrait of Lavoisier and a paragraph about his scientific attainments, including the statement attributed to the presiding judge at his trial: "La République n'a pas besoin de Savans, ni de chymistes" ("The republic has no need of men of science, and not of chemists").

The COW POCK _ or _ the Wonderful Effects of the New Inoculation! _ Vide. the Publications of ŷ Anti-Vaccine Society.

343. A caricature of vaccination. Edward Jenner discovered that immunity to smallpox can be produced by giving a patient the milder disease of cowpox. He based this new practice on observations of dairy maids who frequently became ill with cowpox but never afterward came down with smallpox. At first Jenner's idea was ridiculed, as in this cartoon, because of the mistaken belief that vaccination would actually spread smallpox and also because of people's understandable reluctance to be given an "unnatural" disease. Vaccination eventually came to be accepted after the publication of Jenner's *Inquiry into the Causes and Effects of the Variolae Vaccinae* in 1798.

344. "Give us back our eleven days."
This cartoon by William Hogarth (1754), from "The Election" series, exhibits the popular feeling against the reform of the calendar in 1752, when England decided to change from the old Julian calendar to the new Gregorian calendar. Since there was a difference of eleven days between the two calendrical systems, it was decreed that at the time of changeover, 3 September should be followed on the next day by 14 September, causing simpleminded and ignorant people to believe that they had been robbed of eleven days.

345. The climax of a mesmerist's séance. This satirical cartoon pokes fun at the mesmeric craze that swept through Paris in the 1770s. There is depicted a mesmerist session, centering about the tub (or *bacquet*) that was alleged to store and dispense the invisible fluid that, according to Franz Anton Mesmer, flowed through all humans and was responsible for "animal magnetism." Mesmer contended that some diseases blocked this flow and that an external source of fluid could effect a cure and restore harmony to the body. The fluid was supposedly transmitted from the tub by iron bars and ropes to patients who would undergo convulsions and faint, a stage in the curing process. Mesmer himself appears at the right side of the cartoon; he is depicted with an ass's head, indicating that he is a charlatan. Mesmer drew his theories from various occultists and from respectable sources such as Newton. Compared to blood-letting techniques of the time, his forceful doctor-patient relationship was appealing, and his flair for publicity made him rich in spite of rejection from scientific quarters. Mesmerism as a movement outlasted its originator and was popular well into the nineteenth century, laying the foundation for the hypnotist school, which was to have some influence on Sigmund Freud's development of psychoanalysis.

346. Magnetism unveiled. This eighteenth-century cartoon pictures the mesmerists being confounded by the investigations of a French royal commission of scientists and doctors, whose members included Benjamin Franklin, Lavoisier, astronomer Jean-Sylvain Bailly, Dr. J. I. Guillotin (inventor of the guillotine), and others. Mesmerism seemed a dangerous movement because it offered a radical challenge to established science and religion and even had an admixture of radical political ideas. Franklin is holding up the commission's official report while the mesmerists fly off on broomsticks, with Mesmer portrayed as an ass. The report denied the existence of a mesmeric fluid and stated that all of the effects produced by mesmerists were caused "by the touches of the operator, the excited imagination of the patient, and by the involuntary instinct of imitation."

347. The original orrery. Constructed by John Rowley of London in 1713, or perhaps even earlier, this machine (now on view in the Science Museum, South Kensington, London) was based on one invented by George Graham. Graham had constructed his machine with the help of his uncle, Thomas Tompion, the famous clockmaker. Rowley's orrery came into the possession of Charles Boyle, Fourth Earl of Orrery. The machine's name was established after Sir Richard Steele, the famous essayist, wrote that Rowley "calls his Machine the *Orrery*, in Gratitude to the Nobleman of that Title." This original orrery is much simpler than later ones, and shows only the relative motions of the earth and moon with respect to the sun. Later orreries were constructed to exhibit the motions of all the planets and even their moons, and thus to simulate the operation of the Copernican system. Some orreries were activated by a crank, but others were more complex and relied on clockwork.

348. A Benjamin Martin orrery. This orrery, made entirely of brass and driven by clockwork, was obtained by Harvard College in the 1760s. It closely resembles the instrument seen in Illustration 349 and is geared so that the planets attached to concentric rings in the base complete revolutions around the central sun in times that are proportional to the planets' actual revolutions. Benjamin Martin was a lecturer-demonstrator of experimental philosophy and a noted popularizer of science.

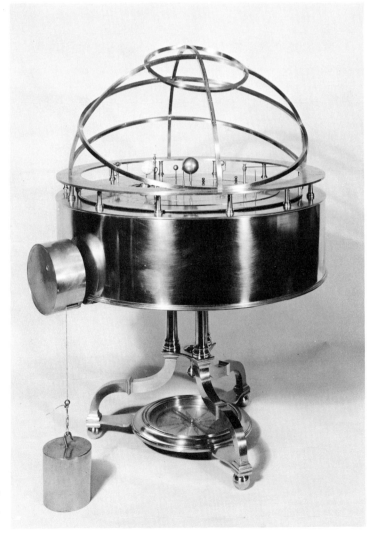

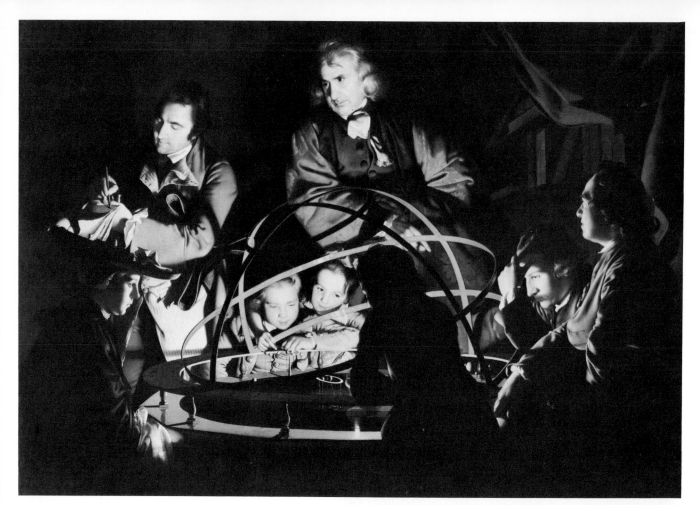

349. A painting of an orrery. In the late eighteenth century, Joseph Wright of Derby made this painting of a group of fascinated spectators, adults and children, watching a demonstration of an orrery. The man at the left is busy taking notes, while the lecturer explains the principles of the celestial motions.

350. An English instrument-maker's trade card of the mid-eighteenth century. As science became a subject of instruction, research, and entertainment, the making of scientific apparatus developed into a new profession and industry. This trade card advertised the instruments made and sold by Edward Scarlett, Junior, "Optician to His Majesty King George the Second." Among the instruments made or sold by Scarlett were spectacles, a magic lantern (*upper right*), a lodestone (shown under the magic lantern), telescope, barometer (*lower right*), prism, drawing instruments, and a Wilson screw-barrel microscope (*lower left*) mounted on a small three-legged stand. Since selling scientific instruments had become an international business, Scarlett published the text of his trade card in French, Dutch, and English.

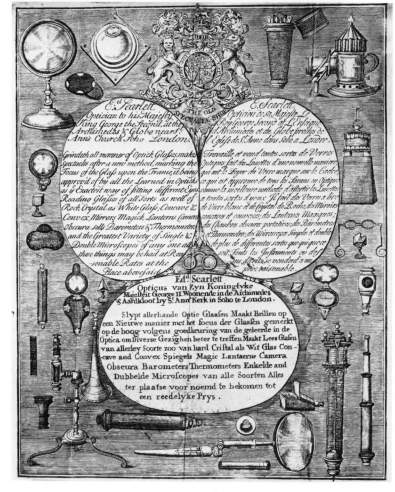

351. A French instrument-maker's trade card of the mid-eighteenth century. "Le Sʳ. Magny" advertises his specialties on his trade cards as "horology and instruments for mathematics and physics, and for the science of mechanics." The card shows cranes and pulleys, a telescope, clocks, an organ, a large armillary sphere, a small universal ring, a barometer, sundials, surveying instruments, mirrors, and an air pump.

352. Large globe of the Russian Academy of Sciences. This huge globe, 3⅓ meters (approximately 11 feet) in diameter, was made between 1748 and 1752 by two mechanics, Benjamin Scott and Filip Tirjutin of the St. Petersburg Academy of Sciences. A map of the earth envelops the exterior surface while the interior contains a representation of the starry heavens. Several viewers can enter the globe through a door in order to sit inside and study the surrounding northern and southern constellations.

353. The constellations. These are representations of the constellations of the northern hemisphere that appear on the interior of the large globe of the St. Petersburg Academy of Sciences.

261

354. Allegorical print celebrating the international cooperation in observing the transit of Venus. The caption reads: "In the midst of a costly war, the King [Louis XVI] was willing to concern himself with the advancement of the sciences. The Academy [of Sciences], represented by the figure of Urania, gives him an account of the transit of Venus across the sun, and other moral and physical observations made on this occasion." Scientists from many nations banded together so that observations of the transit (the passage of Venus across the face of the sun) could be made at stations widely distributed over the earth. The importance of this relatively rare astronomical event (two such transits occur in fairly rapid succession about once in every century) is that it yields an accurate value of the sun's distance from the earth, the fundamental yardstick of our universe.

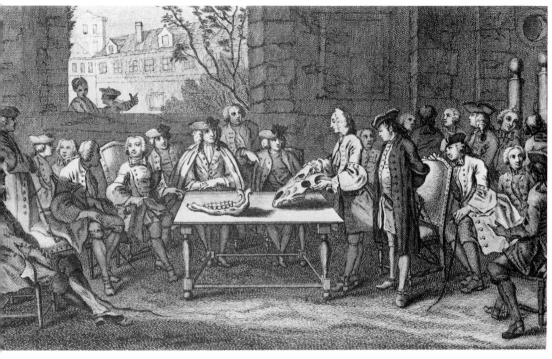

355. Learning about a horse's skull. In a sumptuous book on horsemanship, dated 1733, François Robichon de la Guérinière included a section on the biology of the horse. This illustration from the book shows a group of gentlemen being taught the anatomical features of a horse's skull.

The FIRST Lecture in EXPERIMENTAL PHILOSOPHY.

Engrav'd for the Universal Magazine, for I. Hinton at the Kings. Arms, in S! Pauls Church Yard London 1748.

356. A lecture in experimental philosophy. The *Universal Magazine* for 1748 included this illustration of a group of young men being instructed in experimental physical science. The subject of the lecture, the first in the series, is chemistry. On a shelf at the back of the room are a telescope and other optical instruments to be used in later lectures.

357. Abbé Nollet's demonstration apparatus. In the middle of the eighteenth century, Abbé Nollet published a number of books on performing scientific experiments and demonstrating the fundamental principles of physical science. He performed scientific experiments for the education and delectation of ladies and gentlemen of the French court, as seen in this engraving from his *Lessons in Experimental Physical Science* (1748), a six-volume work that was printed in several editions. The wall cabinets contain apparatus for performing experiments in optics, heat, chemistry, magnetism, statics, and other aspects of science.

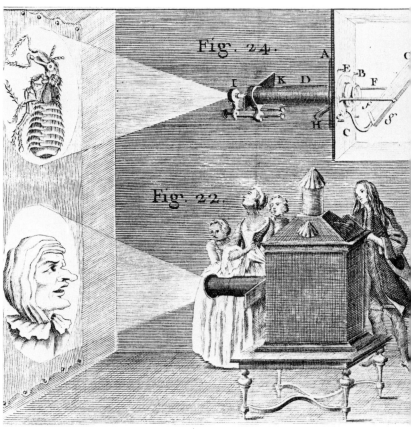

358. Projections with a magic lantern and solar miscroscope. In this print from Abbé Nollett's *Lessons in Experimental Physical Science*, a group of people is staring in amazement at a horrible face (drawn on a slide) projected by a magic lantern and an insect (an actual mounted specimen) projected by a solar microscope. Both instruments were popular sources of scientific as well as nonscientific entertainment. The magic lantern is said to have been the invention of Father Athanasius Kircher, a German Jesuit.

359. Announcement of experiments on electricity. This broadside from Newport, Rhode Island, is dated 16 March 1752 and announces two lectures and demonstrations to be given by Ebenezer Kinnersley, an associate of Benjamin Franklin. Kinnersley gave these lectures in various cities, and, as stated here, performed not only "the most curious" of the experiments "that have been made and published in *Europe*, but a considerable Number of new Ones lately made in *Philadelphia*; to be accompanied with methodical LECTURES on the Nature and Properties of that wonderful Element [Electrical FIRE]."

Newport, March 16. 1752.

Notice is hereby given to the Curious,

That at the COURT-HOUSE, in the Council-Chamber, is now to be exhibited, and continued from Day to Day, for a Week or two;

A COURSE of EXPERIMENTS, on the newly-discovered

Electrical FIRE:

Containing, not only the most curious of those that have been made and published in *Europe*, but a considerable Number of new Ones lately made in *Philadelphia*; to be accompanied with methodical LECTURES on the Nature and Properties of that wonderful Element.

By Ebenezer Kinnersley.

LECTURE I.

I. OF Electricity in General, giving some Account of the Discovery of it.
II. That the Electric Fire is a real Element, and different from those heretofore known and named, and collected out of other Matter (not created) by the Friction of Glass, &c.
III. That it is an extreamly subtile Fluid.
IV. That it doth not take up any perceptible Time in passing thro' large Portions of Space.
V. That it is intimately mixed with the Substance of all the other Fluids and Solids of our Globe.
VI. That our Bodies at all Times contain enough of it to set a House on Fire.
VII. That tho' it will fire inflammable Matters, itself has no sensible Heat.
VIII. That it differs from common Matter, in this; its Parts do not mutually attract, but mutually repel each other.
IX. That it is strongly attracted by all other Matter.
X. An artificial Spider, animated by the Electric Fire, so as to act like a live One.
XI. A Shower of Sand, which rises again as fast as it falls.
XII. That common Matter in the Form of Points attracts this Fire more strongly than in any other Form.
XIII. A Leaf of the most weighty of Metals suspended in the Air, as is said of *Mahomet's* Tomb.
XIV. An Appearance like Fishes swimming in the Air.
XV. That this Fire will live in Water, a River not being sufficient to quench the smallest Spark of it.
XVI. A Representation of the Sensitive Plant.
XVII. A Representation of the seven Planets, shewing a probable Cause of their keeping their due Distances from each other, and from the Sun in the Center.
XVIII. The Salute repulsed by the Ladies Fire; or Fire darting from a Ladies Lips, so that she may defy any Person to salute her.
XIX. Eight musical Bells rung by an electrified Phial of Water.
XX. A Battery of eleven Guns discharged by Fire issuing out of a Person's Finger.

LECTURE II.

I. A Description and Explanation of Mr. Muschenbroek's wonderful Bottle.
II. The amazing Force of the Electric Fire in passing thro' a Number of Bodies at the same Instant.
III. An Electric Mine sprung.
IV. Electrified Money, which scarce any Body will take when offer'd to them.
V. A Piece of Money drawn out of a Person's Mouth in spite of his Teeth, yet without touching it, or offering him the least Violence.
VI. Spirits kindled by Fire darting from a Lady's Eyes (without a Metaphor).
VII. Various Representations of Lightning, the Cause and Effects of which will be explained by a more probable Hypothesis than has hitherto appeared, and some useful Instructions given, how to avoid the Danger of it: How to secure Houses, Ships, &c. from being hurt by its destructive Violence.
VIII. The Force of the Electric Spark, making a fair Hole thro' a Quire of Paper.
IX. Metal melted by it (tho' without any Heat) in less than a thousandth Part of a Minute.
X. Animals killed by it instantaneously.
XI. Air issuing out of a Bladder set on Fire by a Spark from a Person's Finger, and burning like a Volcano.
XII. A few Drops of electrified cold Water let fall on a Person's Hand, supplying him with Fire sufficient to kindle a burning Flame with one of the Fingers of his other Hand.
XIII. A Sulphurous Vapour kindled into Flame by Fire issuing out of a cold Apple.
XIV. A curious Machine acting by means of the Electric Fire, and playing Variety of Tunes on eight musical Bells.
XV. A Battery of eleven Guns discharged by a Spark, after it has passed through ten Foot of Water.

As the Knowledge of Nature tends to enlarge the human Mind, and give us more noble, more grand, and exalted Ideas of the Author of Nature, and if well pursu'd, seldom fails producing something useful to Man, 'tis hoped these Lectures may be tho't worthy of Regard & Encouragement.

*** Tickets to be had at the House of the Widow Allen, in Thames Street, next Door to Mr. John Tweedy's. Price Thirty Shillings each Lecture. The Lectures to begin each Day precisely at Three o'Clock in the Afternoon.

OWERTVRE DV COVRS.

360. A class in chemistry. This woodcut of the beginning of a chemistry course in seventeenth-century France is remarkable because the lecturer has no books or notes, and the students are engaged in a dialogue with their teacher. In the background, an artisan or "lab assistant" is tending the fires.

361. A chemistry lecture in England. This late eighteenth-century colored lithograph by Thomas Rowlandson shows a crowd of spectators watching a demonstration of chemical experiments. The audience is made up of men and women, young and old, who are jammed into the hall, and the lecture table is covered with chemical apparatus of all sorts. As part of the demonstration, the lecturer is pouring a liquid from an ordinary teapot into a glass vial. A small sign over the door behind the demonstrator reads "Surry Institution."

362. Allegorical frontispiece of the *Encyclopédie*. This illustration epitomizes the spirit of the *Encyclopédie* and the Age of Reason, or Enlightenment. According to the description given in the 1751 edition, Truth is shown in the upper central part of the illustration, "wrapped in a veil, radiant with a light which parts the clouds and disperses them." On her left, Reason, wearing a crown, is lifting the veil from Truth, while Philosophy is "pulling it away." At the feet of Truth, "Theology on her knees, receives light from on high." She holds the Bible in her left hand. Immediately below these figures are Geometry, holding a scroll with the theorem of Pythagoras drawn on it; Physics, with her right hand on an air pump; and Astronomy. Below them and to the right are Optics, with a microscope and mirror; Botany, holding a plant; Chemistry, with a retort and furnace; and Agriculture, at the lower right-hand corner. At the right center stand Memory and Ancient and Modern History; History is shown "writing the annals" while—barely discernible—Father Time with his scythe is supporting her book. At the upper left, to the left of Truth, "we see Imagination, who is preparing to adorn and crown Truth." The group immediately underneath Imagination represents the "different genres of poetry (Epic, Dramatic, Satiric, and Pastoral)." Underneath that group and toward the center are "the other Arts of Imitation," Music, Painting, Sculpture, and Architecture. According to the *Encyclopédie*, truth (especially scientific truth) is not a mere abstraction, but a reality with practical effects for man and society, just as the sciences have practical effects on industry and agriculture. This theme is represented by the crowd of practitioners or artisans at the bottom of the illustration, looking upward to scientific truth as their source of progress.

27

The Encyclopédie of Diderot

The *Encyclopédie*, edited by Denis Diderot, began in 1746 as an enterprise of a group of publishers who wished to prepare a French edition of an English encyclopedic dictionary of the sciences. In Diderot's editorial hands, this project was soon transformed into an original work of vast scope and learning, one of the noblest collective enterprises of the Enlightenment. The *Encyclopédie* remained a dictionary in its alphabetical arrangement, but it was also an encyclopedia containing lengthy and authoritative articles on all aspects of knowledge: history and philosophy, theology and ethics, and above all, science and technology. The first volume appeared in 1751; eventually there were seventeen folio volumes of text and eleven supplementary volumes of illustrations. The title page declared that this was an *Encyclopaedia, or Ordered Dictionary of Sciences, Arts, and Trades.*

Each article was written by a specialist, under the general editorial supervision of Diderot. Jean Le Rond d'Alembert, one of the greatest mathematicians of the eighteenth century, was primarily responsible for the articles dealing with mathematics and the exact physical sciences. D'Alembert also wrote the introduction, which stated a general philosophy of science and knowledge, based on the principles of Francis Bacon and lauding John Locke and Isaac Newton. According to d'Alembert, the *Encyclopédie* was not a simple reference work, but a "sanctuary where man's knowledge is protected from time and from revolutions."

The *Encyclopédie* was far from a passive repository of knowledge. It espoused an active philosophy of progress and of the betterment of man's condition on earth. Progress for man and society was seen as a kind of mirror image of the progress that could be seen in the advancement of the sciences. In eighteenth-century France, this was a dangerous political heresy, but Diderot and his associates got by the censors because of the importance of the technical information that their volumes contained. As historian of science Charles C. Gillispie has remarked, "The technology carried the ideology along with it." Eventually, of course, the pressures of reactionary theologians and conservative political authorities were too great for Diderot, and the final volumes were emasculated to satisfy the demands of the censors.

Today the *Encyclopédie* stands as a huge archaeological monument of the scientific and intellectual currents of the Enlightenment; it is rarely read except by scholars engaged in historical research. But the illustrations that accompany the volumes of text are as much a source of delight to today's viewers as they were to those who were fortunate enough to obtain them when they were published. Diderot's goal was to provide accurate records of what was going on in the sciences and in the arts and trades of his day. He maintained that his records would be so perfect in their combination of graphic and verbal information that historians of the future would not need to do research in order to learn about conditions of life in the eighteenth century.

The *Encyclopédie* records how clocks were made, how weapons were forged, how armies and navies were organized and assembled for combat, how chemical experiments were performed, and how men and animals are formed. It explains the laws of motion, the basic principles of the life sciences, and the operation of machines. Art and beauty are discussed, and answers are given to almost any question one can think of about art, history, philosophy, science, or technology. The *Encyclopédie* is thus the sum of all the knowledge of its time, and it is also the expression of man's hope for a better world. That hope was based on a belief in the power of truth—not the truth of revelation, but the truth of nature as revealed by science through experiment and observation and the application of reason to experience. That hope was abetted by the conviction that truth is so great and powerful that it must prevail and must necessarily ensure progress.

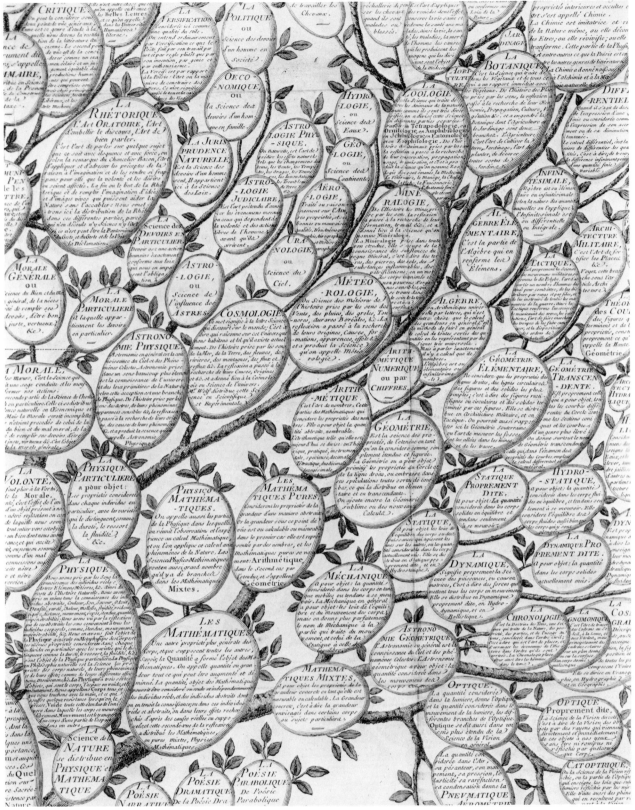

363. The tree of knowledge. This detail of the tree of knowledge from the *Encyclopédie* depicts some of its classifications. As outlined by d'Alembert, understanding in the system of human knowledge falls into three categories: memory, imagination, and reason. Under memory he placed history (both natural history and sacred history), as well as the arts, trades, and manufacturing. Imagination led to poetry, sacred and profane. The central core was reason, where general metaphysics and the "science of God" led to philosophy, the "science of man," and the "science of nature." The science of man embraced ethics and logic, while that of nature led to mathematics and also to "physics," as seen at the lower left. Mathematics was either pure (for instance, arithmetic and geometry) or mixed (comprising mechanics, astronomy, optics, and other branches of ordinary physics). "Particular physics" embraced philology, physical astronomy (and astrology), meteorology, cosmology, botany, zoology, mineralogy, and chemistry. The other branches can be traced on the tree in a similar fashion.

364. The Giant's Causeway. One of the natural wonders pictured in the *Encyclopédie* was the extraordinary collection of basaltic columns in Northern Ireland known as the Giant's Causeway. The columns were found to be part of a continuous underwater geological formation, the other end of which surfaces in Fingal's Cave on the island of Staffa, off the coast of Scotland. This realistic illustration is typical of representations of geology and natural history in the *Encyclopédie*.

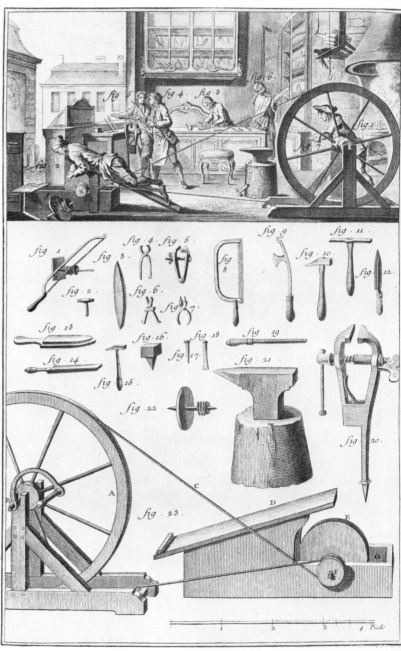

365. The cutlery shop. A plate from the *Encyclopédie* shows various aspects of a cutlery shop and the tools of the cutler. At the top left of the plate, an artisan is shown lying on an inclined board sharpening a knife on a grindstone, which is being turned by a cord looped around a large wheel at right. Diderot was fastidious about having his artists make technically accurate drawings that would show how machines really worked and would enable anyone to construct them. This was very difficult since artists were not used to such exactitude and often did not understand the processes they were delineating. Diderot was particularly concerned with the accuracy of this shop, since his father had been a master cutler. The graphic representation of technological processes was a major feature of the *Encyclopédie*.

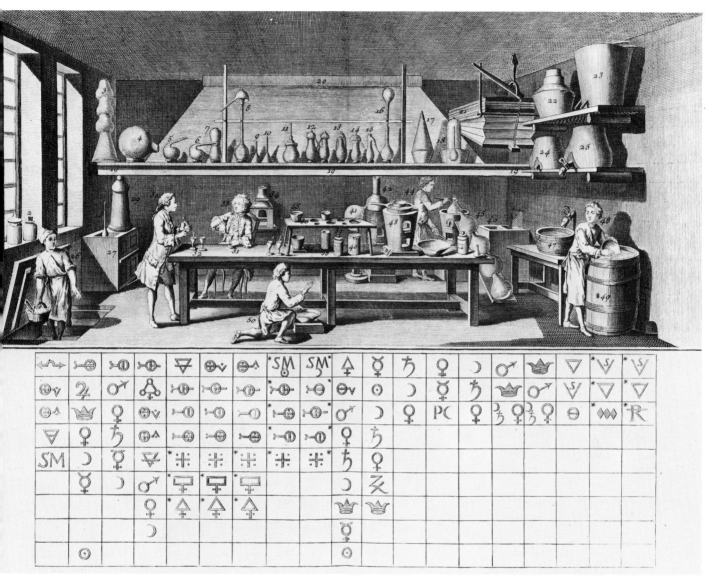

366. Chemistry in the *Encyclopédie*. An engraving of a chemical laboratory shows the different kinds of equipment then in use. The lower half shows a "table of affinities," according to the older system of chemical nomenclature, in which a special symbol was used to designate each chemical substance. This system was replaced in the early nineteenth century by one using letters or combinations of letters, in the style introduced by J. J. Berzelius and still in use today.

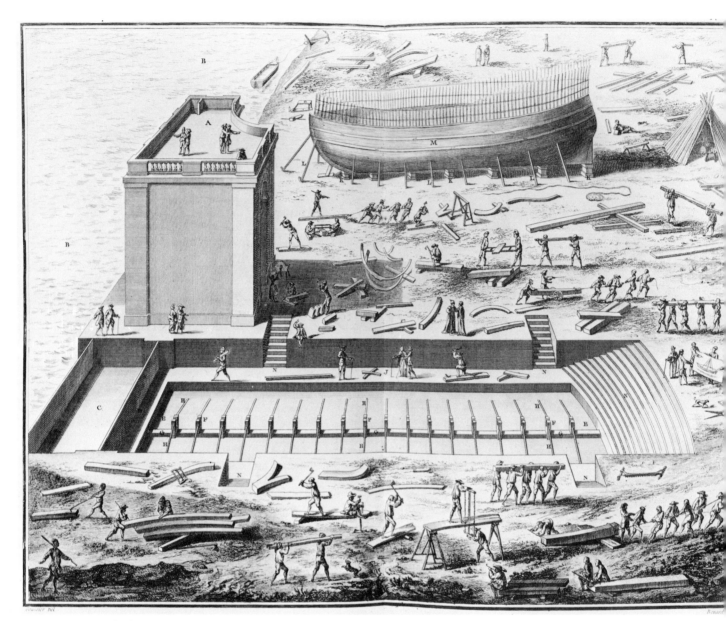

367. A shipyard. Some illustrations in the *Encyclopédie*, like this one of a dry dock and a ship under construction, were large double plates that folded over.

CHAPTER

28

The Belief in Progress and the Limitlessness of Science

The eighteenth century is usually called the Age of Reason or the Enlightenment, in token of the prevailing general belief that the power of the human mind would lead mankind out of the darkness of the past and present into the bright light of the future. There was a commitment to the idea of progress, a continuing process of improving man's conditions of life, his social organization, and even his health. The idea of progress or improvement did not originate in the eighteenth century, but has roots that go back through the seventeenth century to the Renaissance, and even to some degree to the Middle Ages. Isaac Newton was invoking a medieval image of progress, used in the twelfth century by John of Salisbury, when he said that the reason he had been able to see further than his predecessors was that he was standing "on the sholders of Giants."

In the eighteenth century, the notion of progress took on a new dimension that came from the sciences. The scientific intellectual of the eighteenth century, as the historian of science Roger Hahn has put it, felt himself

responsible for the rest of society. His self-confidence and zeal inevitably led him to assume the air of a conqueror and missionary. He was certain the artisan had to be taught to rationalize his activities, the doctor to understand his diagnosis, the musician to mathematicize his art, and the ruler to lead his nation according to natural law. In the metaphoric language of the Enlightenment, the *philosophe* was prepared to lead his nation according to natural law. night with the torch of reason. With his insatiable desire to regenerate and reform all of mankind, the intellectual pictured himself as the true leader of society, placing himself in an elite group within society at large.

In the steady advance in the understanding and control of nature wrought by the sciences, men and women everywhere could see a model of the way in which all kinds of knowledge might be improved, while in the organization of science in nonhierarchical societies, there was provided a model of the way all social activity might be reordered on a more rational basis. The scientific endeavor

368. No limit to science. Francis Bacon, the first major spokesman for the "new science," published his *Instauratio magna* or *Great Instauration* in 1620. It contained the *Novum organum*, which he described as the "True Directions Concerning the Interpretation of Nature." The ship, prominently displayed on the title page of the book, is seen to be sailing through the Pillars of Hercules at the western end of the Mediterranean, which traditionally represented the boundary of man's possible exploration. Here is an expression of the optimistic point of view that knowledge is limitless; in effect, Bacon's emblematic title page declares *"Plus ultra"* ("Yet further"). The Latin quotation at the foot of the waves was taken from the Book of Daniel and reads: "Many will pass through and knowledge will be increased."

thus seemed to show both how knowledge could be increased and how men and women could live and work together in harmony for a general good.

It is sometimes difficult for us today to appreciate the degree of optimism that characterized this general belief in progress. The Marquis de Condorcet, in his *Sketch for a Historical Picture of the Progress of the Human Mind,* written in 1793, during the "1st month of the Year Two of the French Republic," concludes with "hopes for the future condition of the human race . . . under three important heads: the abolition of inequality between nations, the progress of equality within each nation, and the true perfection of mankind." Condorcet looked forward to the day when preventive medicine would improve and when "food and housing becoming healthier" it would produce "a way of life . . . that will develop our physical powers by exercise without ruining them by excess," when "the average length of human life will be increased and better health and a stronger physical constitution will be ensured." Such an "improvement of medical practice" would become "more efficacious with the progress of reason and of the social order" and would result in "the end of infectious and hereditary diseases and illnesses brought on by climate, food, or working conditions." He envisaged a time "when death will be due only to extraordinary accidents or to the decay of the vital forces," so that "the average span between birth and decay will have no assignable value."

Condorcet saw the sciences not merely as the model for progress in society at large but as the fount of new inventions and new modes of preventing disease and increasing health; he was also convinced that there would be a similar improvement in man's "intellectual and moral faculties." Benjamin Franklin expressed similar hopes in a letter to Joseph Priestley in February 1780:

The rapid Progress *true* Science now makes occasions my regretting sometimes that I was born so soon. It is impossible to imagine the Height to which may be carried, in a thousand years, the Power of Man over Matter. We may perhaps learn to deprive large Masses of their Gravity, and give them absolute Levity, for the sake of easy Transport. Agriculture may diminish its Labour and double its Produce; all Diseases may by sure means be prevented or cured, not excepting even that of Old Age, and our Lives lengthened at pleasure even beyond the antediluvian Standard. O that moral Science were in as fair a way of Improvement, that Men would cease to be Wolves to one another, and that human Beings would at length learn what they now improperly call Humanity!

Priestley himself had no doubts about the betterment of the world. As a result of science, he wrote, "the human powers will, in fact, be enlarged." Nature, "including both its materials, and its laws, will be more at our command; men will make their situation in this world more easy and comfortable; they will probably prolong their existence in it." Even more important to Priestley was the firm expectation that "men will grow daily more happy, each in himself, and more able (and, I believe, more disposed) to communicate happiness to others." In short, "whatever was the beginning of this world, the end will be glorious and paradisiacal, beyond what our imagination can now conceive."

In the eighteenth century, however, there were some thinkers who did not fully believe that the continuing progress of the sciences would be shared by man and society. An example is the philosopher David Hume, who supposed that "the world" is neither eternal nor incorruptible, but is probably mortal. Accordingly, he conceived that man himself as a species would share the fate of the world, that is, participate in the decline of "the world" or its "old age." There was no way of telling, he argued in his *Essay on the Populousness of Ancient Nations* (1752), as to whether at the present time man was still advancing toward his highest point of perfection or was declining from it. There were others, like the French economist A. R. J. Turgot, who believed in progress but nevertheless held it necessary to differentiate natural phenomena from the history of man and institutions. Natural phenomena, he held, are subject to uniform and constant laws, operating within "a circle of unchanging revolutions," but in the history of man's activities and his institutions, there is rather a kind of endless and unrepetitive variation.

The thinking men and women of the eighteenth century who believed in progress, and who sought to implement their ideals by social action, were universally committed to a belief in knowledge, and particularly scientific knowledge, as a powerful force for making men and women free and for giving to men and women the power to assume a just and equal position in society. Obviously, this point of view demanded an improvement in the system of education, so that knowledge itself and the ability to use it could be transmitted to the people at large. The commitment to universal and continued progress was also linked to a parallel belief in the continued improvements of the sciences and of the practical arts, especially the arts which would use inventions based on science. Scientists and nonscientists were in agreement that advances in science would provide better control of the environment, as exemplified by Franklin's lightning rod. Who could doubt that Newton's physics was better than Aristotle's, or that Harvey's physiology was superior to Galen's? Improvements were sure to come in every domain of human action: transportation, communication, manufacturing, agriculture, relations between one nation and another, and the organization of society.

Although the historian Edward Gibbon devoted a lifetime to describing and analyzing the *Decline and Fall of the Roman Empire*, he did not believe that our own civilization would be threatened by a similar fate. He said that there are sound reasons to "acquiesce in the pleasing conclusion that every age of the world has increased, and still increases, the real wealth, the happiness, the knowledge and perhaps the virtue of the human race."

Scientists, philosophers, and social reformers generally held that the knowledge of science is limitless. Condorcet stressed the fact that "No-one has ever believed that the mind can gain knowledge of all the facts of nature or attain the ultimate means of precision in the measurement, or in the analysis of the facts of nature, the relations between objects, and all the possible combinations of ideas." It is true, he wrote, that there will always be a part of nature, "always indeed the larger part of it that will remain forever unknown." But he also held that the instruments man's intelligence devises for exploring and understanding the natural world "will increase and improve, the language that fixes and determines his ideas will acquire greater breadth and precision and, unlike mechanics where an increase of force means a decrease of speed, the methods that lead genius to the discovery of truth increase at once the force and the speed of its operations."

For those who linked the progress of man and of society with the sciences, Francis Bacon was an important symbolic figure. Not only had he devoted himself to an exposition of the method of the sciences, but he had also been one of the first prophets of the doctrine that the applications of science to human affairs would result in the betterment of the conditions of life. Furthermore, he had been a pioneer in conceiving how the effectiveness of science would be increased by the organization of scholars into academies and societies. The greatest propagandistic document of the

Enlightenment, the *Encyclopédie*, was conceived and executed under the sign of Bacon and organized according to Baconian ideas. Many of those who believed in science and progress must have had in mind the image of the navigator, directing the forward path of a ship into unknown seas under the sure control provided by the science of navigation. Such a ship is depicted on the title page of Bacon's major work, the *Novum organum*, his exposition of the true method for the advancement of the sciences, the "new logic" of induction.

The men and women of the eighteenth century were well aware that the age in which they lived had witnessed spectacular achievements in science, including the elaboration of the Newtonian scientific revolution, the discovery of the sexual reproduction of flowering plants, the founding of a new branch of science (electricity), and the establishment of modern chemistry in the chemical revolution. As a result of scientific education, as Condorcet put it, "At the present time a young man on leaving school may know more of the principles of mathematics than Newton ever learnt in years of study or discovered by dint of genius." It is little wonder that there should have been a faith in reason, exemplified in its highest manifestation in the exact and natural sciences; that there should have been a hope that society would attain the same rational order discovered in the world of nature; and that men and nations would live together in harmony in a new and better world.

Bibliography and Guide to Further Reading

For those who wish to learn more about the topics presented in this volume, there is a variety of books and monographs available. In the list that follows, I have indicated some of the works that can be helpful to the reader, and I have included some comments where necessary. In many cases it is difficult to choose among a number of first-rate similar works; accordingly, the omission of any book from this list should not be interpreted as a sign of disapproval. I have not attempted to give a complete listing of hardback and paperback reprints for every work, but I have included such reprints as have come to my attention.

1. General Works

There are a great many books, articles, and monographs on the development of science in the sixteenth, seventeenth, and eighteenth centuries, and the number is constantly growing; and yet there are few single general works on science during these centuries that can be wholly recommended for scientific accuracy and true historical perspective. Among these few are Herbert Butterfield's *The Origins of Modern Science, 1300–1800* (New York: Macmillan, 1951; 2nd ed., 1957; a revised printing is in preparation) and A. Rupert Hall's *The Scientific Revolution, 1500–1800: The Formation of the Modern Scientific Attitude* (London, New York: Longmans, Green and Co., 1954; paperback reprint, Boston: Beacon Press, 1970). For the first part of the period discussed in this volume,

Marie Boas's *The Scientific Renaissance, 1450–1630* (New York: Harper & Row, 1962; paperback reprint, New York: Harper Torchbooks, 1966) can be highly recommended. On seventeenth-century science, see Richard S. Westfall's *The Construction of Modern Science: Mechanisms and Mechanics* (New York: Wiley, 1971; reissued by Cambridge Univ. Press [History of Science Series], 1978) and, for an overall view, Charles C. Gillispie's *The Edge of Objectivity: An Essay in the History of Scientific Ideas* (Princeton: Princeton Univ. Press, 1960). Of this group the books by Butterfield and Boas are most suited to the general reader; those by Hall and Gillispie contain the most factual information; the one by Westfall is the most incisive.

The early part of the period under discussion is well presented in A. C. Crombie's *Augustine to Galileo* (2 vols., Harmondsworth [Middlesex, England], Baltimore [Md.]: Penguin, 1969—reprint of revised ed. [1959] of a 1952 work), which is also available under the title *Medieval and Early Modern Science* (2 vols., Garden City [New York]: Doubleday [Anchor Books], 1959), and which may be supplemented by Edward Grant's *Physical Science in the Middle Ages* (New York, London: Wiley, 1971; reissued by Cambridge Univ. Press [History of Science Series], 1978). Allen Debus's *Man and Nature in the Renaissance* (Cambridge [England], London, New York: Cambridge Univ. Press [History of Science Series], 1978) is a brief introduction for the general reader. The essays by Alexandre Koyré, collected under the title *Metaphysics and Measurement: Essays in Scientific Revolution* (London: Chapman & Hall; Cambridge [Mass.]: Harvard Univ. Press, 1968), are challenging and stimulating.

In a class by itself is Robert K. Merton's now-classic *Science, Technology and Society in Seventeenth-Century England*, first published in the journal *Osiris* in 1938 (New York: Fertig; New York: Harper & Row [Harper Torchbooks], 1970; reprinted, New Jersey: Humanities; Sussex [England]: Harvester, 1978). See also the essays on science and religion in seventeenth-century England by S. F. Mason, H. F. Kearney, Christopher Hill, Theodore K. Rabb, and Barbara J. Shapiro in the volume edited by Charles Webster, *The Intellectual Revolution of the Seventeenth Century* (London, Boston: Routledge & Kegan [Past and Present Series], 1974). Reijer Hooykaas's *Religion and the Rise of Modern Science* (Edinburgh, London: Scottish Academic Press, 1972), can be highly recommended.

2. The Nature of the Scientific Revolution and of Revolutions in Science

Thomas S. Kuhn has conceived an original analysis of the concept of revolution that has won much approval and has also aroused criticism; see the second edition of his *The Structure of Scientific Revolutions* (Chicago: Univ. of Chicago Press, 1970—International Encyclopedia of Unified Science, vol. 2, no. 1) and his *The Essential Tension: Selected Studies in Scientific Tradition and Change* (Chicago, London: Univ. of Chicago Press, 1978). For a critique of Kuhn's position, see the volume edited by Imre Lakatos and Alan Musgrave, *Criticism and the Growth of Knowledge* (Cambridge [England]: Cambridge Univ. Press, 1970). A general historical review of the concept of revolutions in science is given in I. B. Cohen's *Revolution in Science: The History, Analysis, and Significance of a Name and a Concept* (Cambridge [Mass.]: Harvard Univ. Press, forthcoming in 1981).

3. Scientific Illustrations and the Relations of Science and Art

On scientific illustrations, see Robert Herrlinger's *History of Medical Illustration from Antiquity to A.D. 1600* (London: Pitman Medical & Scientific Publishing, 1970), David M. Knight's *Natural Science Books in English, 1600–1900* (New York: Praeger, 1972), and Wilfrid Blunt's *The Art of Botanical Illustration* (London: Collins, 1950; New York: Scribner,

1951). Two works that begin the exploration of the role of illustrations in the development of a particular science are Stuart Piggott's *Antiquity Depicted: Aspects of Archaeological Illustration* (London: Thames & Hudson, 1979) and Martin Rudwick's stimulating essay "The Emergence of a Visual Language for Geological Science 1760–1840," in the quarterly journal *History of Science*, vol. 14 (Sept. 1976), pp. 149–195.

On art and science, see Erwin Panofsky's *Galileo as a Critic of the Arts* (The Hague: Martinus Nijhoff, 1954; revised reprint in *Isis*, vol. 47, 1956, pp. 3–15), which is a classic of its kind. See also Panofsky's *Meaning in the Visual Arts* (Garden City [New York]: Doubleday [Anchor Books], 1955), especially pp. 55–107 on "The History of the Theory of Human Proportions as a Reflection of the History of Styles." Additionally, see Sir Ernst Gombrich's *Mirror and Map: Theories of Pictorial Representation* (London: Royal Society, 1975, and in *Philosophical Transactions of the Royal Society*, vol. 270B (1975), pp. 119–149), and his *Art and Illusion: A Study in the Psychology of Pictorial Representation* (New York: Pantheon, 1960). There is much to be learned from the section on science and art in Fritz Saxl's monumental two-volume *Lectures* (London: Warburg Institute, 1957), and his *A Heritage of Images: A Selection of Lectures*, edited by Hugh Honour and John Fleming (Harmondsworth [Middlesex, England], Baltimore: Penguin, 1970), notably "Continuity and Variations in the Meaning of Images" (pp. 13–26) and "The Revival of Late Antique Astrology" (pp. 27–42). An important and suggestive essay by W. M. Ivins, Jr., on the artist and the anatomist may be found in Samuel W. Lambert, Willy Wiegand, and William M. Ivins, Jr.'s *Three Vesalian Essays to Accompany the Icones Anatomicae of 1934* (New York: Macmillan [History of Medicine Series, Library of the New York Academy of Medicine], 1952).

On perspective, see the books by William M. Ivins, Jr., *On the Rationalization of Sight, with an Examination of Three Renaissance Texts on Perspective* (New York: Da Capo, 1938) and *Art and Geometry: A Study in Space Intuitions* (Cambridge [Mass.], Harvard Univ. Press, 1946), plus the important monograph by Samuel Y. Edgerton, Jr., *The Renaissance Rediscovery of Linear Perspective* (New York: Basic, 1975).

See also Geoffrey Lapage's *Art and the Scientist* (Bristol [England]: John Wright & Sons, 1961), Daniel Pedoe's *Geometry and the Liberal Arts* (Harmondsworth [Middlesex, England], Baltimore: Penguin, 1976), and Jacob Opper's *Science and the Arts: A Study in Relationships from 1600–1900* (Rutherford [New Jersey]: Fairleigh Dickinson, 1973).

4. Astronomy

In astronomy, two books by J. L. E. Dreyer remain outstanding—his *History of the Planetary Systems from Thales to Kepler* (Cambridge [England]: Cambridge Univ. Press, 1906), which was reprinted as *A History of Astronomy from Thales to Kepler* (New York: Dover, 1953), and his *Tycho Brahe, a Picture of Scientific Life and Work in the Sixteenth Century* (Edinburgh: Adam & Charles Black, 1890; reprint, New York: Peter Smith, 1972). Another older but splendid book is Francis R. Johnson's *Astronomical Thought in Renaissance England: A Study of the English Scientific Writings from 1500 to 1645* (Baltimore: Johns Hopkins, 1937; reprint, New York: Octagon, 1968). Of major interest is Alexandre Koyré's *The Astronomical Revolution: Copernicus, Kepler, Borelli* (Paris: Hermann; London: Methuen; Ithaca: Cornell Univ. Press, 1973), and his *From the Closed World to the Infinite Universe* (Baltimore: Johns Hopkins, 1957; reprint, 1968).

See also Max Caspar's *Kepler*, trans. and ed. by C. Doris Hellman (London, New York: Abelard-Schuman, 1959), Thomas S. Kuhn's *The Copernican Revolution: Planetary Astronomy in the Development of Western Thought* (Cambridge [Mass.]: Harvard Univ. Press, 1957; reprint, 1968), and Michael Hoskin's *William Herschel and the Construction of the Heavens* (London: Oldbourne, 1963). A popular and readable work is Gale E. Christianson's *The Wild Abyss, the Story of the Men Who Made Modern Astronomy* (New York: Free Press; London: Collier-Macmillan, 1979), with chapters on Copernicus, Tycho Brahe, Kepler, Galileo, and Newton. S. H. Henninger, Jr., has published a work of special value because of its illustrations: *The Cosmographical Glass: Renaissance Diagrams of the Universe* (San Marino [California]: Huntington Library, 1977). See also Harry Woolf's *The Transits of Venus: A Study of Eighteenth-Century Science* (Princeton Univ. Press, 1959; reprint, New York: Arno, forthcoming in 1981).

5. Mathematics

The most recent general account is Morris Kline's *Mathematical Thought from Ancient to Modern Times* (New York: Oxford Univ. Press, 1972). See also Carl B. Boyer's *A History of Mathematics* (New York: Wiley, 1968), and his *The Concepts of the Calculus: A Critical and Historical Discussion of the Derivative and the Integral* (New York: Columbia Univ. Press, 1939; later reprints by other publishers). A short but remarkable introduction to the subject is Dirk J. Struik's two-volume *A Concise History of Mathematics* (New York: Dover, 1948; rev. ed. in 1 vol., 1967). For those trained in mathematics, there is

nothing to compare with D. T. Whiteside's edition (to be completed in 8 vols.) of *The Mathematical Papers of Isaac Newton* (vol. 1, Cambridge [England]: Cambridge Univ. Press, 1967) and his two-volume *The Mathematical Works of Isaac Newton* (New York, London: Johnson Reprint Corp., 1964–1967). See also D. J. Struik's *A Source Book in Mathematics 1200–1800* (Cambridge [Mass.]: Harvard Univ. Press, 1969).

6. Science of Mechanics

A major work is Richard S. Westfall's *Force in Newton's Physics, the Science of Dynamics in the Seventeenth Century* (London: Macdonald; New York: American Elsevier, 1971), which may be supplemented by René Dugas's *Mechanics in the Seventeenth Century* (Neuchâtel [Switzerland]: Editions du Griffon; New York: Central Book Co., 1958). Ernst Mach's *The Science of Mechanics: A Critical and Historical Account of Its Development*, translated by Thomas J. McCormack (6th ed., with revisions through the 9th German ed., LaSalle [Illinois], London: Open Court Publishing Co., 1960) is a philosophical classic, still well worth reading. For a general view of the rise of dynamics in the seventeenth century in relation to astronomy, see I. B. Cohen *The Birth of a New Physics* (Garden City [New York]: Doubleday [Anchor Books], 1960 and later printings; also London: Heinemann, 1961 and later printings), which is written for the general reader.

7. Physics

There is no sound and readable work devoted to the history of physics in this volume's period, although there are many good books on special aspects of physics. For electricity, see John L. Heilbron's insightful and encyclopedic *Electricity in the 17 & 18th Centuries: A Study of Early Modern Physics* (Berkeley [Los Angeles], London: Univ. of California Press, 1979), which sets the subject in the context of a developing profession of physics. On Franklin's theories and experiments in the context of eighteenth-century Newtonian physics, see I. B. Cohen's *Benjamin Franklin, Scientist and Statesman* (New York: Scribner, 1972), and *Franklin: An Inquiry into Speculative Newtonian Experimental Science and Franklin's Work in Electricity and Newton* (Philadelphia: American Philosophical Society, 1956; Cambridge [Mass]: Harvard Univ. Press, 1966; revised paperback reprint forthcoming from Harvard Univ. Press in 1981). The subject of the design, use, manufacture, and distribution of electrostatic generators is presented in W. D. Hackmann's *Electricity from Glass: The History of the Frictional Electrical Machine, 1600–1800* (Alphen aan den Rijn [The Netherlands]: Sijthoff & Noordhoff, 1978).

On optics, A. I. Sabra's *Theories of Light, from Descartes to Newton* (London: Oldbourne, New York: American Elsevier, 1967; reprint forthcoming from Arno, 1981) is an excellent introduction to the subject and may be supplemented by Carl B. Boyer's *The Rainbow from Myth to Mathematics* (New York, London: Thomas Yoseloff, 1959) and Vasco Ronchi's *The Nature of Light, an Historical Survey*, translated by V. Barocas (London: Heinemann; Cambridge [Mass.]: Harvard Univ. Press, 1970).

On aspects of the development of aether theories and the work of Francis Hawksbee, see the essays in Henry Guerlac's *Essays and Papers on the History of Science* (Baltimore: Johns Hopkins, 1977). See also Thomas L. Hankins's *Jean d'Alembert: Science and the Enlightenment* (Oxford: Clarendon, 1970), Robert E. Schofield's *Mechanism and Materialism: British Natural Philosophy in an Age of Reason* (Princeton: Princeton Univ. Press, 1970), Richard Olson's *Scottish Philosophy and British Physics, 1750–1800* (Princeton: Princeton Univ. Press, 1975), Arnold Thackray's *Atoms and Powers: An Essay on Newtonian Matter-Theory and the Development of Chemistry* (Cambridge [Mass.]: Harvard Univ. Press, 1970).

On heat, see Douglas McKie and Niels H. de V. Heathcote's *The Discovery of Specific and Latent Heats* (London: Edward Arnold & Co., 1935; reprint, New York: Arno, 1975).

8. Alchemy and Hermeticism

Two excellent introductions to alchemy are Eric Holmyard's *Alchemy* (Harmondsworth [Middlesex, England], Baltimore: Penguin, 1957 and later reprints) and John Read's *Prelude to Chemistry: An Outline of Alchemy, its Literature and Relationships* (London: G. Bell & Sons, 1936; New York: Macmillan, 1937).

On Hermeticism in relation to modern science, the works of Frances Yates are fundamental; see her classic *Giordano Bruno and the Hermetic Tradition* (London: Routledge & Kegan; Chicago: Univ. of Chicago Press, 1964 and later printings). Miss Yates has summarized her position in her essay in the volume edited by Charles S. Singleton, *Art, Science, and History in the Renaissance* (Baltimore: John Hopkins, 1967); for a contrary view see the essays by Robert S. Westman and J. E. McGuire, *Hermeticism and the Scientific Revolution: Papers Read at a Clark Library Seminar, March 9, 1974* (Los Angeles: Univ. of California, 1977). For a simpler introduction to the subject, see the short book by Allen Debus, listed above in *1. General Works*.

A good way to learn about alchemy is through Betty Jo Teeter Dobbs's *The Foundations of Newton's*

Alchemy: "The Hunting of the Greene Lyon" (Cambridge [England], New York: Cambridge Univ. Press, 1975). The role of Hermeticism in relation to modern science is discussed by a number of experts in *Reason, Experiment, and Mysticism in the Scientific Revolution*, edited by Maria Luisa Righini Bonelli and William R. Shea (New York: Science History, 1975). See also the book by John Maxson Stillman listed below in *9. Chemistry*.

9. Chemistry

The serious reader interested in the development of chemistry may consult the relevant chapters in J. R. Partington's *A History of Chemistry*, vols. 2, 3 (London: Macmillan; New York: St. Martin, 1961, 1962), which are especially valuable for their surveys of the literature and their bibliographies; more readable is Partington's *Short History of Chemistry*, 3rd ed. (London: Macmillan, 1957). Also recommended (for the eighteenth-century background to modern chemistry) is Aaron J. Ihde's *The Development of Modern Chemistry* (New York: Harper & Row, 1964), to be supplemented by Marie Boas's *Robert Boyle and Seventeenth-Century Chemistry* (Cambridge [England]: Cambridge Univ. Press, 1958), Henry Guerlac's *Antoine-Laurent Lavoisier, Chemist and Revolutionary* (New York: Scribner, 1975), and his *Lavoisier—the Crucial Year: The Background and Origin of His First Experiment in Combustion in 1772* (Ithaca: Cornell Univ. Press, 1961). See also John Maxson Stillman's *The Story of Early Chemistry* (London: Constable, 1924), which was reprinted as *The Story of Alchemy and Early Chemistry* (New York: Dover, 1960), and Ida Freund's *The Study of Chemical Composition: An Account of Its Method and Historical Development, with Illustrative Quotations* (Cambridge [England]: Cambridge Univ. Press, 1904; paperback reprint, New York: Dover, 1968).

10. Earth Science, including Paleontology

Martin J. S. Rudwick's *The Meaning of Fossils: Episodes in the History of Paleontology* (London: Macdonald; New York: American Elsevier, 1972; rev. ed., New York: Science History, 1976) is a readable and authoritative introduction to the subject. Some older general histories of geology covering the period of this volume are Frank Dawson Adams's *The Birth and Development of the Geological Sciences* (Baltimore: Williams & Wilkins, 1938; New York: Dover, 1954), Sir Archibald Geikie's *The Founders of Geology* (London: Macmillan, 1897), and Karl Alfred von Zittel's *History of Geology and Paleontology to the End of the Nineteenth Century* (London: Walter Scott; New York: Scribner, 1901). See also Francis C.

Haber's *The Age of the World, Moses to Darwin* (Baltimore: Johns Hopkins, 1959), and *A Source Book in Geology, 400–1900*, edited by Kirtley F. Mather and Shirley L. Mason (Cambridge [Mass.]: Harvard Univ. Press, 1967), and the volume edited by Cecil J. Schneer, *Toward a History of Geology* (Cambridge [Mass.], London: MIT Press, 1969).

11. Maps and Navigation

A superb introduction to the subject of navigation is David Watkins Waters's *The Art of Navigation in England in Elizabethan and Early Stuart Times* (London: Hollis & Carter; New Haven: Yale Univ. Press, 1958). See also E. G. R. Taylor's *The Haven-Finding Art: A History of Navigation from Odysseus to Captain Cook* (London: Hollis & Carter, 1956) and Charles H. Cotter's *A History of Nautical Astronomy* (London, Sydney, Toronto: Hollis & Carter; New York: American Elsevier, 1968).

On maps and cartography, see Leo Bagrow's *History of Cartography*, revised and enlarged by R. A. Skelton (Cambridge [Mass.]: Harvard Univ. Press, 1964). See also Lloyd A. Brown's *The Story of Maps* (Boston: Little, Brown, 1949; reprint, New York: Dover, 1980), and the catalog issued by the Walters Art Gallery entitled *The World Encompassed, an Exhibition of the History of Maps, Held at the Baltimore Museum of Art, October 7 to November 23, 1952* (published by the trustees of the gallery, 1952), which is profusely illustrated.

On geography, there is an interesting collection assembled by George Kisch, *Source Book in Geography* (Cambridge [Mass.]: Harvard Univ. Press, 1979).

On the marine chronometer, see Rupert Thomas Gould's *The Marine Chronometer: Its History and Development* (London: J. D. Potter, 1923; reprint, Cedar Knolls [N.J.]: Wenman Bros., 1975).

12. The Life Sciences

The most readable general introduction is Erik Nordenskiöld's *The History of Biology: A Survey* (New York: Tudor, 1935 and numerous other editions; reprint, St. Clair Shores [Mich.]: Scholarly, 1976). Julius von Sachs's *History of Botany (1530–1860)* (Oxford: Clarendon Press, 1890; 2nd. ed., 1906; reprint, New York: Russell and Russell, 1967) is a perennial classic and still worth studying. The following studies can be highly recommended: Joseph Needham's *A History of Embryology* (Cambridge [England]: Cambridge Univ. Press, 1934; rev. reprint, 1975; reprint, New York: Arno, 1975), Francis Joseph Cole's *A History of Comparative Anatomy from Aristotle to the Eighteenth Century* (London: Macmillan, 1944; reprint, New York: Dover, and Magnolia [Mass.]: Peter

Smith, 1975), Arthur William Mayer's *The Rise of Embryology* (Stanford: Stanford Univ. Press, 1939; reprint, New York: Arno, forthcoming in 1981), and Elizabeth Gasking's *Investigations into Generation 1651–1828* (Baltimore: John Hopkins, 1967). See also Sir Geoffrey Keynes's *William Harvey* and—for serious readers—Walter Pagel's *Harvey's Biological Ideas*, both listed below in *13. Physiology and Medicine*.

See also Wilfrid Blunt's *The Compleat Naturalist, A Life of Linnaeus*, with a discussion of the Linnaean classification by William T. Stearn (London: Collins; New York: Dutton, 1971), and Knut Hagberg's *Carl Linnaeus*, translated by Alan Blair (London: Cape, 1952), and Richard W. Burkhardt, Jr., *The Spirit of System: Lamarck and Evolutionary Biology* (Cambridge [Mass.]: Harvard Univ. Press, 1977) is an excellent study.

See also Agnes Arber's *Herbals, Their Origin and Evolution* (new ed., Cambridge [England]: Cambridge Univ. Press, 1938), Frank J. Anderson's *An Illustrated History of the Herbals* (New York: Columbia Univ. Press, 1977), F. Dawtrey Drewitt's *The Romance of the Apothecaries' Garden at Chelsea* (London: Chapman and Dodd, 1922; 2nd ed., 1924), and Robert John Harvey-Gibson's *Outlines of the History of Botany* (London: A. & C. Black, 1919; reprint, New York: Arno, forthcoming in 1981). A major reference work is the *Catalogue of Botanical Books in the Collection of Rachel McMasters Miller Hunt* (Pittsburgh: Hunt Botanical Library, 1958–1961). Wilfrid Blunt's *The Art of Botanical Illustration* (London: Collins, 1950) is just as suggestive as it is beautiful to look at. Thomas S. Hall's *A Source Book in Animal Biology* (Cambridge [Mass.]: Harvard Univ. Press, 1951) is a valuable collection of primary documents.

13. Physiology and Medicine

Sir Michael Foster's *Lectures on the History of Physiology During the Sixteenth, Seventeenth, and Eighteenth Centuries* (Cambridge [England]: Cambridge Univ. Press, 1901; reprint, 1961) is a readable introduction, which may be supplemented by John Farquhar Fulton's *Selected Readings in the History of Physiology* (Springfield [Illinois]: Charles C Thomas, 1930; rev. and enlarged ed., prepared by Leonard Wilson, 1966) and his volume in the Clio Medica Series, *Physiology* (New York: Paul B. Hoeber, 1931; reprint, New York: AMS Press, 1978).

Two general works on medical history are recommended: Charles Singer and E. Ashworth Underwood's *A Short History of Medicine*, 2nd ed. (New York, Oxford: Oxford Univ. Press, 1962) and Richard Harrison Shryock's *The Development of Modern Medicine: An Interpretation of the Social and Scien-*

tific Factors Involved (New York: Knopf, 1947; reprint, New York: Hafner, 1969); the Singer-Underwood volume is over 800 pages long.

On seventeenth- and eighteenth-century medicine, three books by Lester King are well-written introductions: *The Philosophy of Medicine: The Early Eighteenth Century* (Cambridge [Mass.]: Harvard Univ. Press, 1977), which is also good for the seventeenth century; *The Road to Medical Enlightenment 1650–1695* (London: Macdonald; New York: American Elsevier, 1970); and, especially, *The Medical World of the Eighteenth Century* (Chicago: Univ. of Chicago Press, 1958; reprint, New York: Robert A. Krieger, 1971).

On the circulation of the blood, a good introduction is Charles Singer's *The Discovery of the Circulation of the Blood* (London: William Dawson [originally published, 1922], New York: Science History Publications, 1956), supplemented by Kenneth D. Keele's *Leonardo da Vinci on Movement of the Heart and Blood* (London: Harvey & Blythe, 1952). See also Walter Pagel's superb and learned study *William Harvey's Biological Ideas: Selected Aspects and Historical Background* (Basel, New York: S. Karger, 1967) and Gweneth Whitteridge's informative and authoritative *William Harvey and the Circulation of the Blood* (London: Macdonald, New York: American Elsevier, 1971). Especially recommended are Sir Geoffrey Keynes's magisterial *The Life of William Harvey* (Oxford: Clarendon Press, 1966), and Kenneth D. Keele's shorter and more easily readable *William Harvey: The Man, the Physician, and the Scientist* (London: Nelson, 1965).

See also June G. Goodfield's *The Growth of Scientific Physiology: Physiological Method and the Mechanist-Vitalist Controversy, Illustrated by the Problems of Respiration and Animal Heat* (London: Hutchinson, 1960; reprint, New York: Arno, 1975), and Everett Mendelsohn's *Heat and Life: The Development of the Theory of Animal Heat* (Cambridge [Mass.]: Harvard Univ. Press, 1964).

14. Anatomy

The major work in this area is F. J. Cole's *History of Comparative Anatomy* listed above in *12. Physiology and Medicine*. A short and readable account is given in Charles Singer's *Evolution of Anatomy: A Short History of Anatomical and Physiological Discovery to Harvey* (New York: Knopf, 1926; reprint, New York: Dover, 1957). A good reference and guide, though somewhat old-fashioned and opinionated, is Johann Ludwig Choulant's *History and Bibliography of Anatomic Illustration in Its Relation to Anatomic Science and the Graphic Arts*, translated and edited by Mortimer Frank (Chicago: Univ. of Chicago Press, 1920; reprint, New York: Schuman, 1945).

See also George W. Corner's *Anatomy* (New York: Paul B. Hoeber, Clio Medica Series, 1930; reprint, New York: Hafner, 1964) and his *Anatomical Texts of the Earlier Middle Ages* (Washington, D.C.: Carnegie Institution, 1927); and Fielding H. Garrison's *The Principles of Anatomic Illustration Before Vesalius; Being an Inquiry into the Rationale of Artistic Anatomy* (New York: Paul B. Hoeber, 1926).

The major work on the subject of anatomical illustrations is by Robert Herrlinger (see *3. Scientific Illustrations . . .*), which ends with 1600; the second part of this work, Marielene Putscher's *Geschichte der Medizinischen Abbildung von 1600 bis zur Gegenwart* (Munich: Heinz Moos Verlag, 1972) has not yet appeared in an English translation.

A rich source of information concerning anatomy and the anatomical theater is William S. Heckscher's *Rembrandt's Anatomy of Dr. Nicolaas Tulp: An Iconological Study* (New York: New York Univ. Press, 1958). See also J. Playfair McMurrich's *Leonardo da Vinci the Anatomist (1452–1519)* (Baltimore: Williams & Wilkins, 1930).

For anatomy in the sixteenth century, there are two collections of texts and illustrations, both a delight to the eye and both edited and translated by Charles D. O'Malley and J. B. de C. M. Saunders: *Leonardo da Vinci on the Human Body* (New York: Henry Schuman, 1952) and *The Illustrations from the Works of Andreas Vesalius of Brussels* (Cleveland, New York: World, 1950).

15. Anthropology and Archaeology

A recent work is Annemarie De Waal Malefijt's *Images of Man: A History of Anthropological Thought* (New York: Knopf, 1974), which may be supplemented by Robert H. Lowie's *The History of Ethnological Theory* (New York: Farrar & Rinehart, 1937). See also Margaret T. Hodgen's *Early Anthropology in the Sixteenth and Seventeenth Centuries* (Philadelphia: Univ. of Pennsylvania Press, 1964; paperback reprint, 1971).

On archaeology, a good introduction is Glyn Daniel's *Man Discovers His Past* (London: Duckworth, 1966; New York: Crown, 1966, 1968), supplemented by C. W. Ceram's *Gods, Graves and Scholars, The Story of Archaeology* (London: Gollancz; New York: Knopf, 1951, 1967; reprint, New York: Bantam, 1976) and Daniel's *A Picture History of Archaeology* (London: Thames & Hudson, 1958).

There are many historical anthologies and readers in archaeology. Among them are Glyn Daniel's *The Origins and Growth of Archaeology* (New York:

Thomas Y. Crowell, 1968) and C. W. Ceram's *The World of Archaeology: The Pioneers Tell Their Own Story* (London: Thames & Hudson, 1966).

Two works that aim to link the origins and development of archaeology and of anthropology are Stanley Casson's *The Discovery of Man: The Story of the Inquiry into Human Origins* (New York, London: Harper & Brothers, 1939) and the volume edited by Peter B. Hammond, *Physical Anthropology and Archaeology: Selected Readings* (New York, London: Macmillan, 1964).

16. Academies and Scientific Societies

Roger Hahn's *The Anatomy of a Scientific Institution: The Paris Academy of Science, 1666–1803* (Berkeley [Los Angeles], London: Univ. of California Press, 1971) is a superb full-dress history of a scientific institution, set in its social matrix with a complete record of its scientific activities. Another highly recommended work is Robert E. Schofield's *The Lunar Society of Birmingham: A Social History of Provincial Science and Industry in Eighteenth-Century England* (Oxford: Clarendon Press, 1963); for a French counterpart, see Maurice Crosland's *The Society of Arcueil: A View of French Science at the Time of Napoleon* (London: Heinemann; Cambridge [Mass.]: Harvard Univ. Press, 1967). For the St. Petersburg Academy, see Alexander Vucinich's *Science in Russian Culture: A History to 1860* (Stanford: Stanford Univ. Press, 1963); for the Royal Society, see Charles Richard Weld's two-volume *A History of the Royal Society* (London: John W. Parker, 1848; reprint, New York: Arno, 1975), and Margery Purver's *The Royal Society: Concept and Creation* (London: Routledge and Kegan; Cambridge [Mass.]: MIT Press, 1967), which can be supplemented by Thomas Sprat's *The History of the Royal Society of London* (London: printed by T. R. for J. Martyn . . ., 1667; facsimile reprint, St. Louis: Washington Univ. Press; London: Routledge and Kegan, 1959, 1966), which contains a record of the society's endeavors and a defense of the aims of the new organization. For the Accademia del Cimento, see W. E. Knowles Middleton's *The Experimenters: A Study of the Accademia del Cimento* (Baltimore: Johns Hopkins, 1972), which can be supplemented by a selection of the academy's research projects, published in Richard Waller's English version as *Essayes of Natural Experiments made in the Academie del Cimento* (London: printed for Benjamin Alsop, 1684; reprinted, with an introduction by A. Rupert Hall, New York, London: Johnson Reprint Corp., 1964). The best general study of academies is still the older work by Martha Ornstein [Bronfenbrenner], *The Role of Scientific Societies in the Seventeenth Cen-* tury (Chicago: Univ. of Chicago Press, 1928, 1938; reprint, New York: Arno, 1975), which can be supplemented by Harcourt Brown's *Scientific Organizations in Seventeenth-Century France (1620–1680)* (Baltimore: Williams & Wilkins, 1934; reprint, New York: Russell & Russell, 1967).

17. Scientific Instruments and Instrument-Makers

On scientific instruments, see Maurice Daumas's *Scientific Instruments of the Seventeenth and Eighteenth Centuries* (Atlantic Highlands [N.J.]: Humanities Press, 1972). Of great interest is H. R. Calvert's *Scientific Trade Cards in the Science Museum Collection* (London: Her Majesty's Stationery Office, 1971).

On instruments for astronomy, navigation, surveying, optics, and sundials, see the beautifully illustrated book by Harriet Wynter and Anthony Turner, *Scientific Instruments* (New York: Scribner, 1975). See also H. C. King's *The History of the Telescope* (Cambridge [Mass.]: Sky Publishing, 1955), to be supplemented by Albert Van Helden, *The Invention of the Telescope* (*Transactions of the American Philosophical Society*, vol. 67 (1977), part 4), containing a collection, in translation, of original documents. The standard reference work on microscopes has long been *The History of the Microscope* by Reginald S. Clay and Thomas S. Court (London: Charles Griffin, 1932; reprint, Kennebunkport [Maine]: Longwood Press, 1977), which should be used in conjunction with G. L.'E. Turner's critical comments in his article, "Micrographia Historica: The Study of the History of the Microscope," in *Proceedings of the Royal Microscopical Society*, vol 7 (1972), pp. 1220–1249. More recent is S. Bradbury's *The Microscope: Past and Present* (Oxford, London, New York: Pergamon, 1968).

See also Edmond R. Kiely's *Surveying Instruments, Their History and Classroom Use* (New York: Bureau of Publications, Teachers College, 1947 [National Council of Teachers of Mathematics, Nineteenth Yearbook]). A guide to an eighteenth-century collection is given in G. L.'E. Turner's *Van Marum's Scientific Instruments in Teyler's Museum: Descriptive Catalogue*, offprinted from vol. 4 of the *Hollandsche Maatschappij der Wetenschappen*'s edition of Martinus van Marum's *Life and Work* (Leyden: Noordhoff International Publishing, 1969–1976). The only available study in depth of an eighteenth-century maker of scientific instruments is John R. Milburn's *Benjamin Martin: Author, Instrument-Maker and "Country Showman"* (Alphen aan den Rijn [The Netherlands]: Sijthoff & Noordhoff, 1976). On electrical instruments, see the book by W. D. Hackmann listed above in *7. Physics*.

18. Technology and Invention

A good general source of information is to be found in volumes 4 and 5 of the five-volume *History of Technology*, edited by Charles Singer, E. J. Holmyard, A. Rupert Hall, and Trevor I. Williams (Oxford: Clarendon Press, 1954–1958), and the one-volume compilation (and expansion of parts of that work) by T. K. Derry and Trevor I. Williams, *A Short History of Technology* (New York, Oxford: Oxford Univ. Press, 1961). See also Abbott Payson Usher's *A History of Mechanical Inventions* (rev. ed., Cambridge [Mass.]: Harvard Univ. Press, 1954), and Friedrich Klemm's anthology of primary sources, *A History of Western Technology*, translated by Dorothea Waley Singer (London: George Allen & Unwin, 1959).

For technology in the Renaissance and the seventeenth century, see Alexander Gustav Keller, ed., *A Theater of Machines* (London: Chapman & Hall, 1964; New York: Macmillan, 1964, 1965).

For the eighteenth century, see A. E. Musson and Eric Robinson's *Science and Technology in the Industrial Revolution* (Manchester: Manchester Univ. Press, 1969), and A. E. Musson's *Science, Technology, and Economic Growth in the Eighteenth Century* (London: Methuen & Co., 1972). See also David S. Landes's *The Unbound Prometheus: Technological Change and Industrial Development in Western Europe from 1750 to the Present* (Cambridge [England]: Cambridge Univ. Press, 1969).

For special technologies, see P. Butler's *The Origin of Printing in Europe* (Chicago: Univ. of Chicago Press, 1940), and H. G. Aldis's *The Printed Book*, 2nd ed., (Cambridge [England]: Cambridge Univ. Press, 1941). A magisterial study of the printing of a single great work is Robert Darnton's *The Business of Enlightenment: A Publishing History of the Encyclopédie, 1775–1800* (Cambridge [Mass.]: Harvard Univ. Press, 1979); also, Charles C. Gillispie's *A Diderot Pictorial Encyclopedia of Trade and Industry* (New York: Dover, 1959). See also Charles Singer's *The Earliest Chemical Industry* (London: Folio Society, 1948), Frank Sherwood Taylor's *A History of Industrial Chemistry* (New York: Abelard-Schuman, 1957; London: Heinemann, 1957; reprint, New York: Arno, 1972); H. W. Dickinson's *A Short History of the Steam Engine* (Cambridge [England]: Cambridge Univ. Press, 1939); and E. C. Smith's *Short History of Naval and Marine Engineering* (Cambridge [England]: Cambridge Univ. Press, 1938).

19. Classics of Science from Leonardo to Lavoisier

There are many classics of science from the sixteenth, seventeenth, and eighteenth centuries that are available for today's reader. Among them the following are of major interest: Nicholas Copernicus's *On the Revolutions [of the Celestial Spheres]*, edited by Jerzy Dobrzycki, translation and commentary by Edward Rosen (Baltimore: Johns Hopkins, 1978). Copernicus's earlier and shorter *Commentariolus* has been translated with an extensive technical apparatus by Noel Swerdlow, "The Derivation and First Draft of Copernicus's Planetary Theory: A Translation of the *Commentariolus* with Commentary," in *Proceedings of the American Philosophical Society*, vol. 117 (1973), pp. 423–512. See also Edward Rosen's earlier translation of the *Commentariolus* in his *Three Copernican Treatises*, 3rd ed. (New York: Octagon, 1971). For Vesalius and Leonardo da Vinci, see listing above in *14. Anatomy*. See also Jean Paul Richter's illustrated two-volume edition of *The Literary Works of Leonardo da Vinci*, 2nd ed., enlarged and revised by Jean Paul Richter and Irma A. Richter (London, New York, Toronto: Oxford Univ. Press, 1939; the original unrevised two-volume edition has been reprinted, New York: Dutton, 1970).

For the seventeenth century, see William Harvey's *The Circulation of the Blood, and Other Writings*, in the Kenneth J. Franklin translation (Oxford: Blackwell's Scientific Publications [for the Royal College of Physicians of London], 1957; reprint in "Everyman's Library," London: Dent; New York: Dutton, 1963, 1968). Isaac Newton's *Opticks* is available in a paperback reprint (New York: Dover, 1952; corrected reprint, 1979); a new edition has been completed by Henry Guerlac. The *Principia* is currently available in a facsimile reprint of the Andrew Motte translation (1729) and in a revised and modernized version of the Motte translation, prepared by Florian Cajori (Berkeley [Los Angeles]: Univ. of California Press, 1946 and later paperback reprints). A new translation has been completed by I. B. Cohen and Anne Whitman and is to be published by Harvard Univ. Press and Cambridge Univ. Press. Robert Hooke's *Micrographia*, a delight to read and to look at, is available in four different reprints: one, as vol. 13 of *Early Science in Oxford*, ed. by R. T. Gunther (Oxford: printed for the subscribers, 1938); two, as vol. 20 of *Historiae Naturalis Classica*, ed. by J. Cramer and H. K. Swann (New York: Hafner, 1961); three, as a paperback (New York: Dover, 1961); and four, as a hardback reprint (Brussels: Editions Culture et Civilisation, 1966). The "Culture et Civilisation" edition is the best of the four, insofar as doing justice to the plates is concerned.

Blaise Pascal's *The Equilibrium of Liquids and the Weight of the Mass of the Air* is to be found in the translation by I. H. R. Spiers and A. G. H. Spiers in *The Physical Treatises of Pascal* (New York: Columbia

Univ. Press, 1937). Three scientific treatises of Descartes have been translated by Paul J. Olscamp in *Discourse on Method, Optics, Geometry and Meteorology* (Indianapolis, New York, Kansas City: Bobbs-Merrill [Library of Liberal Arts], 1965); the *Discourse on Method* is also available in other translations, among them one by F. E. Sutcliffe, *Discourse on Method and the Meditations* (Harmondsworth [Middlesex, England], New York: Penguin, 1968 and later printings).

See also Francis Bacon's *The Advancement of Learning*, ed. by G. W. Kitchin (London: Dent [Everyman's University Library], 1973; Totowa [New Jersey]: Rowman & Littlefield, 1974 and other printings), and *The New Organon*, ed. by Fulton H. Anderson (Indianapolis, New York: Bobbs-Merrill [Library of Liberal Arts], 1960).

Some eighteenth-century scientific classics include Antoine-Laurent Lavoisier's *Elements of Chemistry, in a New Systematic Order, Containing all the Modern Discoveries*, translated by Robert Kerr (Edinburgh, 1790; paperback reprint, New York: Dover, 1965); René Antoine Ferchault de Réaumur's *The Natural History of Ants*, translated and annotated by William Morton Wheeler (New York: Knopf, 1926); Albrecht von Haller's *First Lines of Physiology*, a reprint of the 1786 English edition (New York, London: Johnson Reprint Corp., 1966); Benjamin Franklin's *Experiments and Observations on Electricity*, a facsimile reprint of the 1st edition (London, 1751, 1753–54), edited by I. B. Cohen (Washington, D.C.: Smithsonian Institution Press [for the Dibner Library, National Museum of History and Technology], forthcoming in 1981); Stephen Hales's *Vegetable Staticks*, with a foreword by Michael A. Hoskin (London: Macdonald; New York: American Elsevier, 1969). A collection, *Mikhail Vasil'evich Lomonosov on the Corpuscular Theory*, has been translated by Henry M. Leicester (Cambridge [Mass.]: Harvard Univ. Press, 1970). See also Antoine-Nicholas de Condorcet's *Sketch for a Historical Picture of the Progress of the Human Mind*, translated by June Barraclough (New York: Noonday Press, 1955), and Jean Le Rond d'Alembert's *Preliminary Discourse to the Encyclopedia of Diderot*, translated by Richard N. Schwab (Indianapolis, New York: Bobbs-Merrill [Library of Liberal Arts], 1936).

20. Some Biographies and Monographs on the Work of Individual Scientists

Walter Pagel's *Paracelsus: An Introduction to Philosophical Medicine in the Era of the Renaissance* (Basel, New York: S. Karger, 1958) and Charles D. O'Malley's *Andreas Vesalius of Brussels, 1514–1564*

(Berkeley [Los Angeles]: Univ. of California Press, 1964) are scholarly works that can be highly recommended. See also J. L. E. Dreyer's biography of Tycho Brahe listed above in *4. Astronomy*.

For Galileo, there is Stillman Drake's magisterial *Galileo at Work: His Scientific Biography* (Chicago, London: Univ. of Chicago Press, 1978), and Ludovico Geymonat's briefer introduction, *Galileo: A Biography and Inquiry into His Philosophy of Science* (New York, Toronto, London: McGraw-Hill, 1965), translated by Drake. See also Pasquale M. D'Elia's *Galileo in China* (Cambridge [Mass.]: Harvard Univ. Press, 1960), Jerome L. Langford's *Galileo, Science, and the Church* (rev. ed., Ann Arbor: Univ. of Michigan Press [Ann Arbor Paperbacks], 1971), and Giorgio de Santillana's *The Crime of Galileo* (Chicago: Univ. of Chicago Press, 1955; paperback reprint, 1955). Alexandre Koyré's *Galileo Studies* (London: Harvester Press, 1978), first published in French in 1939, is a pioneering work that has had an enormous influence on the development of the history of science in the years since World War II.

For Kepler, see Max Caspar's biography, *Kepler* (London, New York: Abelard Schuman, 1959; reprint, New York: Collier Books, 1962), translated and edited by C. Doris Hellman. Paolo Rossi's *Francis Bacon: From Magic to Science* (London: Routledge & Kegan; Chicago: Univ. of Chicago Press, 1968; paperback reprint, 1978) can be highly recommended. See also the biographies of Harvey by Sir Geoffrey Keynes, K. D. Kiele, and Walter Pagel listed above in *13. Physiology and Medicine*.

Some seventeenth-century studies of general interest are Richard S. Westfall's *Science and Religion in Seventeenth-Century England* (New Haven: Yale Univ. Press, 1958; paperback reprint, Ann Arbor: Univ. of Michigan Press, 1973), Marjorie Hope Nicolson's *Newton Demands the Muse: Newton's Opticks and the Eighteenth Century Poets* (Princeton: Princeton Univ. Press, 1946), and the collection of Nicolson's essays on science and the literary imagination, *Science and Imagination* (Ithaca: Cornell Univ. Press —Great Seal Books, 1956, 1962; reprint, Hamden [Conn.]: Archon Books, 1976). For the illustrations made by John White and Jacques Le Moyne of the flora and fauna and the customs and ways of life of the native inhabitants of America, see Stefan Lorant's *The New World: The First Pictures of America* (New York: Duell, Sloan & Pearce, 1946).

Also for the seventeenth-century, see Marie Boas's *Robert Boyle and Seventeenth-Century Chemistry* (Cambridge [England]: Cambridge Univ. Press, 1958), supplemented by her *Robert Boyle on Natural Philosophy: An Essay with Selections from His Writings*

(Bloomington: Indiana Univ. Press, 1965), and R. E. W. Maddison's *The Life of the Honourable Robert Boyle* (London: Taylor & Francis; New York: Barnes & Noble, 1969).

There are innumerable books and monographs on Isaac Newton: Richard S. Westfall's two-volume *"Never at Rest": A Biography of Isaac Newton* (Cambridge [England], New York: Cambridge Univ. Press, [announced for 1980]) is certain to be accurate, interesting, and provoking, while I. B. Cohen's *Isaac Newton: A Life in Science* (New York, London: Scribner [announced for 1981]), based on an extensive account in vol. X of the *Dictionary of Scientific Biography*, deals primarily with Newton's "hard" science. On Newton's alchemy, see Betty Jo Teeter Dobbs's monograph listed above in *8. Alchemy and Hermeticism*. For Newton's mathematics, the pair of introductions to Derek T. Whiteside's two-volume edition of *Mathematical Works of Isaac Newton* (New York, London: Johnson Reprint Corp., 1964, 1967) is a convenient summary. See also I. B. Cohen's *The Newtonian Revolution* (Cambridge [England], New York: Cambridge Univ. Press, 1980). For a stimulating social history of the Newtonian natural philosophy, see Margaret C. Jacob's *The Newtonians, and the English Revolution* (Ithaca: Cornell Univ. Press, 1976).

See also Margaret 'Espinasse's *Robert Hooke* (London: William Heinemann, 1956), Charles E. Raven's *John Ray, Naturalist: His Life and Works* (Cambridge [England]: Cambridge Univ. Press, 1942), and Michael Sean Mahoney's *The Mathematical Career of Pierre de Fermat (1601–1665)* (Princeton: Princeton Univ. Press, 1973).

For the eighteenth century, highly recommended are Thomas L. Hankins's *Jean d'Alembert* (listed above in *7. Physics*), C. Stewart Gillmor's *Coulomb and the Evolution of Physics and Engineering in Eighteenth-Century France* (Princeton: Princeton Univ. Press, 1971), Henry Guerlac's *Lavoisier* (see *9. Chemistry*), and Wilfrid Blunt's *Linnaeus* and Richard Burkhardt's *Lamarck*, both listed in *12. The Life Sciences*. See also Robert E. Schofield's compilation, *A Scientific Autobiography of Joseph Priestley (1733–1804)* (Cambridge [Mass.], London: MIT Press, 1966), Boris N. Menshutkin's *Russia's Lomonosov: Chemist, Courtier, Physicist, Poet* (Princeton: Princeton Univ. Press, 1952), E. St. John Brooks's *Sir Hans Sloane: The Great Collector and His Circle* (London: Batchworth Press, 1954), Knut Hagberg's *Carl Linnaeus*, translated by Alan Blair (London: Cape, 1952), H. C. Cameron's *Sir Joseph Banks, the Autocrat of Philosophers* (London: Batchworth Press, 1952), the collection edited by Eric Robinson and Douglas McKie, *Partners in Science: Letters of James Watt and*

Joseph Black (Cambridge [Mass.]: Harvard Univ. Press, 1970), Archibald Edmund Clark-Kennedy's *Stephen Hales, D.D., F.R.S., an Eighteenth Century Biography* (Cambridge [England]: Cambridge Univ. Press, 1929; reprint, Ridgewood [N.J.]: Gregg Press, 1965), and Gerritt Lindebroom's *Hermann Boerhaave: The Man and His Work* (London: Methuen, 1968). Some studies of general interest are Robert Darnton's witty and illuminating *Mesmerism and the End of the Enlightenment in France* (Cambridge [Mass.]: Harvard Univ. Press, 1968), Philip C. Ritterbush's *Overtures to Biology: The Speculations of Eighteenth-Century Naturalists* (New Haven, London: Yale Univ. Press, 1964), Shirley Roe's *Matter, Life, and Generation: 18th-Century Embryology and the Haller-Woolf Debate* (Cambridge [England], New York: Cambridge Univ. Press, forthcoming in 1981), and Terence Doherty's *The Anatomical Works of George Stubbs* (London: Secker & Warburg, 1974). Also see three works listed above in *7. Physics*: Robert E. Schofield's *Mechanism and Materialism*, Richard Olson's *Scottish Philosophy and British Physics*, and Arnold Thackray's *Atoms and Powers*.

Albert Bettax's *The Discovery of Nature* (New York: Simon & Schuster, 1965) is a well-illustrated, popular presentation of aspects of science during the period from Leonardo to Lavoisier and also during the nineteenth and twentieth centuries.

21. Further Information

The foregoing list contains a selection from among many works that can help a reader to obtain more information. The best way to find out more about the lives and achievements of individual scientists is to consult the *Dictionary of Scientific Biography*, edited by Charles C. Gillispie (16 vols., New York: Scribner, 1970–1980). For an introduction to current literature in the history of science, see the annual "Critical Bibliography," published as part of *Isis* (the quarterly journal of the History of Science Society); these annual bibliographies will be particularly helpful to those wishing to consult the ever-growing periodical literature. An excellent source is Magda Whitrow's *ISIS Cumulative Bibliography: A Bibliography of the History of Science formed from ISIS Critical Bibliographies 1–90*—vols. 1–2 (London: Mansell, 1971) is devoted to writings concerning individual scientists and institutions; vol. 3 (London: Mansell, 1976) is devoted to subjects; and vol. 4 (in progress) will be devoted to historical periods and geographical regions or countries. A useful introductory guide is David Knight's *Sources for the History of Science 1660–1914* (Ithaca: Cornell Univ. Press, 1975).

Picture Sources and Credits

In the following list of picture sources and credits, some of the book titles are abbreviated; in every case, however, there is sufficient information to enable the reader to identify the source unambiguously. It will be observed that in some instances there is a discrepancy between the dates given in the illustration caption and in the following list; the reason is that the caption usually refers to the first edition or the first publication, whereas the actual source of the illustration may be a later edition or printing. If a book or article is fully identified in a caption (by title, in the original language, and by date), the title is not repeated here. If, however, in the caption the title has been given in an English translation, it then appears below in the original language, together with the place and date of publication. In the case of a bibliographical puzzle or rarity, some additional information or a reference to a source of information has been added to the citation; and in some other cases, the information given in the caption has been modified or amplified here.

1. *A new universal history of arts and sciences, showing their origin, progress, theory, use and practice,* London, 1759, vol. 1, frontispiece. Harvard College Library.

2. Coronelli, Marco Vincenzo (or Vincenzo Maria), *Atlante veneto,* Venice, 1691, vol. 1, facing title page.

By permission of the Houghton Library, Harvard University.

3. Rodler, Hieronymus, *Eyn schön nützlich büchlin,* Siemern, 1531. By permission of the Houghton Library, Harvard University.

4. Schön, Erhard, *Unnderweissung der proportzion,* Nuremberg, 1538. By permission of the Houghton Library, Harvard University.

5. Pen and ink drawing in the Windsor Royal Library, Courtauld Institute of Art; studies of the head and shoulders of a man, Windsor Castle Manuscripts, folio a, 2 v, Windsor 19001 (Popham; no. 243). Reproduced by gracious permission of Her Majesty Queen Elizabeth II.

6. Vitruvius Pollio, *De architectura,* Florence, 1522. By permission of the Houghton Library, Harvard University.

7. Pen and ink drawing (Popham; no. 215). Accademia delle Belle Arti, Venice. Photo by Osvaldo Bohm.

8. Oil painting, 1480. Chiesa degli Ognissanti. Charles Eames photo, Office of Charles and Ray Eames, Venice, Calif.

9. Reproduced by courtesy of the Trustees, The National Gallery, London.

10. Woodcut in the Zentralbibliotek, Zurich, Wijkiana Broadsides, vol. 15, no. 5. Photo by Owen Gingerich.

11. William Hayes Fogg Art Museum, Harvard University.

12. Albertina Museum, Vienna.

13. Cinti Chapel, Church of Santa Maria Maggiore, Rome. Photo by Gallerie Pontificie.

14. The Louvre, Paris. French Embassy Press and Information Division.

15. "La nature est le livre des philosophes." I. Bernard Cohen collection.

16. This engraving exists in several states and issues, some reversed. Copyright British Museum.

17. Oil painting in The Orangery, Uppsala, Sweden. Photo courtesy of the Swedish Linnaean Society.

18. Gamelin, Jacques, *Nouveau recueil d'ostéologie et de myologie*, Toulouse, 1779. Philip Hofer collection, Cambridge, Mass.

19. Pastel in the Alte Pinakotek, Munich. Bavarian State Painting Collection, Munich.

20. Drawing. Bibliothèque Nationale, Paris.

21. Fuchs, Leonhart, *De historia stirpium*, Basel, 1542. Burndy Library.

22. Burndy Library.

23. Caius Plinius Secundus, *Historia naturalis*, Venice, 1469. By permission of the Houghton Library, Harvard University.

24. Valturius, R., *Elenchus et index rerum militarium*, Verona, 1472. Burndy Library.

25. Rare Book Division, New York Public Library.

26. Arnold Arboretum, Harvard University.

27. Brunfels, Otto, *Herbarum vivae eicones ad natura imitationem*, Strasbourg, 1530. By permission of the Houghton Library, Harvard University.

28. First page of *Preclarissimus liber elementorum Euclidis, . . . in artem geometrie . . .*, Venice, 1482. There are some copies with a different *incipit*; there is no title page. Burndy Library.

29. Archimedes, *Opera omnia*, Basel, 1544 (Thomas Gechauff Venatorius, editor). By permission of the Houghton Library, Harvard University.

30. Copernicus, Nicolaus, *De revolutionibus orbium coelestium*, Nuremberg, 1543, page 9, verso. Burndy Library.

31. Abraham bar-Ḥiyya, *Sphaera mundi*, Basel, 1546. New York Public Library.

32. *L'Astrologo*, engraving by Giovanni Volpato, after an original by Domenico Maggiotto, Venice, 1770. I. Bernard Cohen collection.

33. Fludd, Robert, *Utriusque cosmi maioris scilicet et minoris metaphysica, physica atque technica historia*, Oppenheim, 1617. By permission of Houghton Library, Harvard University.

34. Manuscript horoscope. By permission of the Houghton Library, Harvard University.

35. Manuscript horoscope, Royal Greenwich Observa-

tory, MSS 18/2ʳ. National Maritime Museum, Greenwich.

36. By permission of the Houghton Library, Harvard University.

37. See note for Illustration 33.

38. [Raleigh, Sir Walter; Hulsius, Levinius], *Kurtze wunderbare Beschreibung*. Des goldzeichen Konigreichs Guianae in America oder newen Welt unter der Linea Aequinoctiali gelegen . . . von dem wolgebornen Herrn Walther Raleigh . . . Jetzt aber ins Hochteutsch gebracht . . . durch Levinum Hulsium, Nuremberg, 1599. Herzog August-Bibliothek, Wolfenbüttel.

39. Léger, Jean, *Histoire générale des églises évangéliques des vallées de Piémont; ou Vaudoises*, Leiden, 1669. Harvard College Library.

40. By permission of the Houghton Library, Harvard University.

41. Khunrath, Heinrich, *Amphitheatrum sapientiae aeternae*, Hanover, 1609. By permission of the Houghton Library, Harvard University.

42. Maier, Michael, *Atalanta fugiens*, Oppenheim, 1617. The Countway Medical Library, Boston.

43. From the reprint in *Musaeum Hermeticum reformatum et amplificatum*, Frankfurt, 1678, page 393. By permission of the Houghton Library, Harvard University.

44. Mylius, Johann Daniel, *Philosophia reformata*, Frankfurt, 1622. The University of Wisconsin Library.

45. Cabinet des Estampes, Bibliothéque Nationale, Paris.

46. Engraved by F. Basan: "*Le Plaisir des Fous*, à Paris chés Basan Graveur rue St. Jacques et à Amsterdam chés P. Fouquet." I. Bernard Cohen collection.

47. Gilded sphere at the Museo di Storia della Scienza, Florence. Charles Eames photo, Office of Charles and Ray Eames, Venice, Calif.

48. By permission of the Houghton Library, Harvard University.

49. Finé, Oronce, holograph manuscript of *Le sphere du monde*, Paris, 1549 (translated by Finé himself from his *De mundi spaera sive cosmographia*, Paris, 1542). By permission of the Houghton Library, Harvard University.

50. The British Library, London.

51. Copernicus, Nicolaus, holograph manuscript of *De revolutionibus orbium coelestium*, Nuremberg, 1543. Jagiellonian Library, Kraków; Charles Eames photo, Office of Charles and Ray Eames, Venice, Calif.

52. Apianus, Petrus, *Cosmographia, per Gemmam Phrysiam . . . restituta*, Antwerp, 1539. By permission of the Houghton Library, Harvard University.

53. Digges, Thomas, *A Perfit Description of the*

Caelestials Orbes, published as an addition to Digges, Leonard, *A Prognostication Everlastinge . . . Lately Corrected and Augmented by Thomas Digges,* London, 1576. By permission of the Houghton Library, Harvard University.

54. Tycho Brahe, *Astronomiae instauratae mechanica,* Wandsbeck, 1598. (The portrait is dated 1597.) Burndy Library.

55. See note for Illustration 54.

56. Blaeu, Willem Janszoon, *Novus atlas,* Amsterdam, 1642–1646. By permission of the Houghton Library, Harvard University.

57. See note for Illustration 54.

58. Kircher, Athanasius, *Iter exstaticum coeleste,* ed. 3, Würzburg, 1671, facing page 37. By permission of the Houghton Library, Harvard University.

59. Riccioli, Giovanni Battista, *Almagestum novum,* Bologna, 1651. By permission of the Houghton Library, Harvard University.

60. Galileo Galilei, *Istoria e dimostrazioni intorno alle macchie solari . . .,* Rome 1613. By permission of the Houghton Library, Harvard University.

61. Museo di Storia della Scienza, Florence. Charles Eames photo, Office of Charles and Ray Eames, Venice, Calif.

62. These drawings appeared as prints in Galileo's *Siderius nuncius,* Venice, 1610. Biblioteca Nazionale Centrale, Florence; photo by Guido Sansoni.

63. Schall von Bell, Johann Adam (or Thang Jo-Wang), *Yuan-ching Shuo,* Peking, 1626. Yen Ching Library, Harvard University.

64. MSS of Lord Egremont-Leconfield; Petworth MSS 241/IX, f. 30. Photo courtesy of Prof. John Shirley.

65. Galileo Galilei, MSS Galileanae P. 3, T. 3, Car. 30r. Biblioteca Nazionale Centrale, Florence; photo by Guido Sansoni.

66. Kepler, Johannes, *Mysterium cosmographicum,* Tübingen, 1597. Burndy Library.

67. Kepler, Johannes, *Astronomia nova,* Prague, 1609. By permission of the Houghton Library, Harvard University.

68. Kepler, Johannes, *Tabulae Rudolphinae,* Ulm, 1627. By permission of the Houghton Library, Harvard University.

69. By permission of the Houghton Library, Harvard University.

70. Cabinet des Estampes, Bibliothèque Nationale, Paris.

71. Huygens, Christiaan, *Systema Saturnium,* The Hague, 1659. By permission of the Houghton Library, Harvard University.

72. By permission of the Houghton Library, Harvard University.

73. William H. Schab Gallery, Inc.; photo by Eric Pollitzer.

74. Fontenelle, Bernard Le Bouyer [or Bovier] de, *Entretiens sur la pluralité des mondes,* in *Oeuvres diverses,* vol. 1, The Hague, 1728. By permission of the Houghton Library, Harvard University.

75. MS in the State Historical Museum, Moscow.

76. Harvard College Library.

77. Halley, Edmond, *A description of the passage of the shadow of the moon over England . . .,* London, 1715. By permission of the Houghton Library, Harvard University.

78. See note for Illustration 77.

79. *Etat du ciel pendant l'éclipse totale du soleil qui doit arriver à Paris le 22 mai 1724.* Annotated broadside in the Cabinet des Estampes, Bibliothèque Nationale, Paris.

80. Engraving by David Herrliberger. Department of Prints and Engravings, Zentralbibliothek, Zurich.

81. Engraving by J. G. Kaid, 1771, after a portrait from the life by W. Pohl. I. Bernard Cohen collection.

82. Euler, Leonhard, *Theoria motuum planetarum et cometarum . . .,* Berlin, 1744. Courtesy of the Humanities Research Center Library, The University of Texas at Austin.

83. Bode, Johann Elert, *Uranographia sive astrorum descriptio,* Berlin, 1801. Harvard College Library.

84. Herschel, William, "Account of Some Observations Tending to Investigate the Construction of the Heavens," in *Philosophical Transactions of the Royal Society,* vol. 74 (1784), pp. 437–451. Royal Society, London; photo by John R. Freeman & Co.

85. Hevelius, Johannes, *Machinae coelestis pars prior,* Danzig (now Gdansk), 1673, fig. M, facing p. 222. By permission of the Houghton Library, Harvard University.

86. See note for Illustration 85, fig. EE, facing p. 444.

87. Joh. Mich. Franz's Collection of Engraved Mappae Astronomicae in the Niedersächsische Staats und Universitäts Bibliothek, Göttingen.

88. Horrebow, Peder Nielson, *Basis astronomiae,* Copenhagen, 1735. The British Library, London.

89. Huygens, Christiaan, *Horologium oscillatorium,* Paris, 1673. By permission of the Houghton Library, Harvard University.

90. University of Basel Library.

91. Etching by Francis Place (probably in 1676), after a drawing by Robert Thacker. From the complete set of prints in the Pepysian Library, Magdalene College, Cambridge. (See Howse, Derek, *Francis Place and the Early History of the Greenwich Observatory,* New York, 1975.) Photo courtesy of Derek Howse.

92. I. Bernard Cohen collection.

93. Bibliothèque Nationale, Paris; Photo Viollet.

94. Le Comte, Louis, *Nouveaux mémoires sur l'état présent de la Chine*, Paris, 1696. By permission of the Houghton Library, Harvard University.

95. Copper engraving by Johann Adam Delsenbach, entitled *Das Nürnbergl. Observatorium Astronom. wie es von dem Seel. Hn. G: C: Eimmart berühmten Mathem. aufgerichtet worden*, Nuremberg, 1716. Photo by Owen Gingerich from the Crawford Collection, Royal Observatory in Edinburgh; used by permission of the Astronomer Royal of Scotland.

96. Archive of the Soviet (USSR) Academy of Sciences in Leningrad.

97. Harvard College Library.

98. This copper engraving by J. Walker was issued with a caption saying it was published by William Herschel, 1 February 1794. It also appeared in Herschel, William, "Description of a Forty-feet Reflecting Telescope," in *Philosophical Transactions of the Royal Society*, vol. 85 (1795), pp. 347–409. Science Museum, South Kensington, London.

99. Reisch, Gregorius, *Margarita philosophica*, Freiberg, 1503. By permission of the Houghton Library, Harvard University.

100. Stöffler, Johannes, *Elucidatio fabricae ususque astrolabii*, Tübingen, 1511. By permission of the Houghton Library, Harvard University.

101. Ruff, Walter, *Der . . . mathematischen und mechanischen Künst . . .*, Nuremberg, 1547. By permission of the Houghton Library, Harvard University.

102. Napier, John, *Logarithmorum canonis descriptio seu arithmeticarum supputationum . . .*, London, 1620. By permission of the Houghton Library, Harvard University.

103. By permission of the Houghton Library, Harvard University.

104. From a MS letter from Schickard to Kepler, 25 February 1624; original in Leningrad. Photo courtesy of Dr. Martha List, Kepler Kommission, Munich.

105. One of Pascal's original machines; at present in the IBM Historical Collection. Photo courtesy of International Business Machines Corp., Armonk, N.Y.

106. *Machines et inventions approuvées par l'Académie royale des sciences*, Paris, 1735, tome 4. By permission of the Houghton Library, Harvard University.

107. Leibniz's original machine; in the Niederssächsische Landesbibliothek, Hannover.

108. By permission of the Houghton Library, Harvard University.

109. Leibniz, G. W. von, "Nova methodus pro maximis et minimim . . .," in *Acta eruditorum*, vol. 3 (Oct. 1684), p. 467.

110. University Library, Cambridge, England, MS Add. 4004, fol. 81 verso.

111. Newton, Isaac, *Opticks*, 1704, p. [170] of the second numeration. By permission of the Houghton Library, Harvard University.

112. Stevin, Simon, *De beghinselen der weeghconst*, Leiden, 1586. By permission of the Houghton Library, Harvard University.

113. MS copy of *Dialogo del Galileo* [*delle nuove scienze*], MS gal. 78, fol. 534. Biblioteca Nazionale Centrale, Florence.

114. By permission of the Museum of Comparative Zoology; on deposit in the Houghton Library, Harvard University.

115. Varignon, Pierre, *Projet d'une nouvelle méchanique*, Paris, 1687, p. 89. By permission of the Houghton Library, Harvard University.

116. Delmedigo, Joseph Solomon, *Elim . . .*, Amsterdam, 1629. Harvard College Library Hebrew Collection.

117. By permission of the Houghton Library, Harvard University.

118. Plate 3 for vol. 3; in vol. 5 (plates and index). I. Bernard Cohen collection.

119. Coulomb, Charles Augustin de, *Théorie des machines simple*, nouvelle edition, Paris, 1821 (after the Paris 1781 edition). Harvard College Library.

120. Huygens, Christian, "De motu corporum ex percussione," tab. 1, facing p. 80 of Huygens's *Opera reliqua*, vol. 2, contained in Huygens's *Opuscula posthuma*, Amsterdam, 1728. By permission of the Houghton Library, Harvard University.

121. Book 2, p. 56. By permission of the Houghton Library, Harvard University.

122. Marci, Johannes Marcus, *De proportione motus*, Prague, 1639. The British Library, London.

123. See note for Illustration 122.

124. Varignon, Pierre, *Nouvelles conjectures sur la pesanteur*, Paris, 1690. By permission of the Houghton Library, Harvard University.

125. Münster, Sebastian, *Rudimenta mathematica*, Basel, 1551. By permission of the Houghton Library, Harvard University.

126. Musschenbroek, Petrus van, *Cours de physique expérimentale et mathématique*, trans. by Sigaud de la Fond, Paris 1769, vol. 1, plate 18. The original Latin edition, *Introductio ad philosophiam naturalem*, ed. by J. Lulofs, was published (posthumously) in Leiden in 1762. Harvard College Library.

127. Photo courtesy of the Royal Society of London.

128. MS Add. 4004, fol. 10 verso. University Library, Cambridge, England.

129. Burndy Library.

130. P. 402 of Newton's own interleaved (and annotated) copy of the first edition of the *Principia*, Adv. b. 39.1. University Library, Cambridge, England.

131. Newton, Isaac, *A Treatise of the System of the World*, London, 1728. University Library, Cambridge, England.

132. One of a set of eighteenth-century devices constructed under the auspices of the Grand Duke Pietro Leopold di Lorena, intended to illustrate the principles of mechanics. Museo di Storia della Scienza, Florence.

133. Engraved by James Basire, drawn by T. Malton. Harvard College Library.

134. Broadside in the Goldsmith Company's Library, University College Library, London.

135. *Allegorical Monument to Isaac Newton*, oil painting (1727–1730) on canvas (220x139cm) by Giovanni Battista Pittoni, with portions painted by Domenico and Giuseppe Valeriani. Reproduced by courtesy of the Fitzwilliam Museum, Cambridge, England.

136. Witelo, *Optica*, Nuremberg, 1535. (A second edition or printing, 1551; then again in a volume edited by Friedrich Risner, with a Latin translation of Ibn Al-Haytham's treatise on optics, under the general *Opticae thesaurus*, Basel, 1572. Witelo's treatise should be called *Perspectiva*.

137. Meyer, Cornelius (or Meijer, Cornelis), *Nuovi ritrovamenti*, Rome, 1689. By permission of the Houghton Library, Harvard University.

138. Descartes, René, *Dioptrique*, 1637. By permission of the Houghton Library, Harvard University.

139. 'sGravesande, Wilhelm Jacob, *Physices elementa mathematica experimentis confirmata*, Leiden, 1722. By permission of the Houghton Library, Harvard University.

140. MS sketch of Newton's "experimentum crucis," 1721. Jeffery Elkins Collection of Newton MSS, Bodleian Library, Oxford University.

141. Table 1 in Book 2. By permission of the Houghton Library, Harvard University.

142. Huygens, Christiaan, *Traité de la lumiere*, Leiden, 1690. From the facsimile reprint, Brussels, 1967.

143. MS E, p. 93. The Dutch Society of Sciences.

144. By permission of the Houghton Library, Harvard University.

145. The Science Museum, South Kensington, London.

146. Dobell, Clifford, *Antony van Leeuwenhoek and His "Little Animals,"* London-Amsterdam, 1932.

147. Hooke, Robert, *Micrographia*, London, 1665. Burndy Library.

148. By permission of the Houghton Library, Harvard University.

149. Microscope made by Pieter Lyonet. The Science Museum, South Kensington, London.

150. Harvard College Library.

151. Nollet, Jean-Antoine, *Leçons de physique expérimentale*, Paris, 1748, 6 vols. (It is not stated specifically in the text that this pair is actually a mother and daughter.) Harvard College Library.

152. Appendix 2, plate 2, p. 368. Photo courtesy of David P. Wheatland, Collection of Historical Scientific Instruments, Harvard University.

153. Photo courtesy of David P. Wheatland, Collection of Historical Scientific Instruments, Harvard University.

154. In the Science Museum, South Kensington, London. British Crown Copyright.

155. A lodestone from the Urals in a brass mounting. State Historical Museum, Moscow.

156. Gilbert, William, *De magnete*, London, 1600. Burndy Library.

157. Schott, Gaspar, *Technica curiosa*, Würzburg, 1664, plate 12, opp. p. 203. By permission of the Houghton Library, Harvard University.

158. Guericke, Otto von, *Experimenta nova Mageburgica de vacuo spatio*, Amsterdam, 1672. By permission of the Houghton Library, Harvard University.

159. See note for Illustration 158.

160. Boyle, Robert, *A Continuation of New Experiments Physico-Mechanical Touching the Spring and Weight of the Air*, Oxford, 1669, plate 5. By permission of the Houghton Library, Harvard University.

161. *Universal Magazine*, vol. 4 (1749), facing p. 311. Harvard College Library.

162. Bossut, Charles, *Nouvelles expériences sur la résistance des fluides*, Paris, 1777. Harvard College Library.

163. Lavoisier, Antoine-Laurent, *Traité élémentaire de chimie*, Paris, 1789. By permission of the Houghton Library, Harvard University.

164. *Philosophical Transactions of the Royal Society*, vol. 90 (1800), part 2, plate 11. Harvard College Library.

165. Desaguliers, Jean Theophilus, *A Course of Experimental Philosophy*, London, 1744, vol. 2. By permission of the Houghton Library, Harvard University.

166. Farey, John, *A Treatise on the Steam Engine*, London, 1827, plate 11. Harvard College Library.

167. By permission of the Houghton Library, Harvard University.

168. Burndy Library.

169. Watson, William, *Expériences et observations pour servir à l'explication de la nature et des proprietes de l'électricité*, Paris, 1748, plate 3. Burndy Library.

170. Nollet, Jean-Antoine, *Recherches sur les causes*

particulières des phenomènes électriques, Paris, 1764, 5th Disc., plate 2. (The first edition was published in Paris in 1749, the second in 1753.) By permission of the Houghton Library, Harvard University.

171. Harvard College Library.

172. Burndy Library.

173. By permission of the Houghton Library, Harvard University.

174. MS belonging to the American Academy of Arts and Sciences, currently kept in the Athenaeum, Boston. Photo by the Office of Charles and Ray Eames, Venice, Calif.

175. Beck, Dominicus, *Kurzer Entwurf der Lehre von der Elecktricität*, Salzburg, 1787, from the title page. Photo courtesy of David P. Wheatland, Collection of Historical Scientific Instruments, Harvard University.

176. M. V. Lomonosov Museum, Leningrad.

177. Chappe d'Auteroche, Jean-Baptiste, "Expérience sur l'électricité naturelle," separate printing of plate that appears in *Voyage en Sibérie, fait par ordre du roi en 1761*, Paris, 1768. I. Bernard Cohen collection.

178. Marum, Martinus van, *Description d'une très grande machine électrique*, Haarlem, 1785, plate 1. Harvard College Library.

179. Coulomb, Charles Augustin de, *Construction et usage d'une balance électrique*, Paris, 1785, plate 13. Burndy Library.

180. Drawing made for Galvani's *De viribus electricitatis in motu musculari*, Bologna, 1791. From the reproductions of the wash drawings in the archives of the Science Academy of the Bologna Institute, published by Battaglia-Rangoni S.p.A. in Bologna.

181. *Philosophical Transactions of the Royal Society*, vol. 90 (1800), pp. 403–431, plate 17. Burndy Library.

182. Oil painting at Rockefeller University, New York City. Photo Bulloz.

183. Brunschwig, Hieronymus, *Das Buch der rechten Kunst zu Distillieren*, Strasbourg, 1500. Burndy Library.

184. Ashmole, Elias, *Theatrum chemicum Britannicum*, London, 1652, p. 51.

185. The Countway Medical Library, Boston.

186. Becher, J. J., *Tripus hermeticus*, Frankfurt, 1689. Harvard College Library.

187. Derby Museum and Art Gallery, Derby, England.

188. Cabinet des Estampes, Bibliothèque Nationale, Paris.

189. Dodart, Denis, *Mémoires pour servir à l'histoire des plantes*, Paris, 1676. By permission of the Houghton Library, Harvard University.

190. Harvard College Library.

191. Cabinet des Estampes, Bibliothèque Nationale, Paris.

192. Priestley, Joseph, *Experiments and Observations on Different Kinds of Air*, London, 1774, vol. 1, plate 1. By permission of the Houghton Library, Harvard University.

193. See note for Illustration 192, plate 2.

194. Lavoisier, Antoine-Laurent, *Elements of Chemistry in a New Systematic Order*, translated by Robert Kerr, Edinburgh, 1790. By permission of the Houghton Library, Harvard University.

195. Lavoisier, Antoine-Laurent, *Traité élémentaire de chimie*, Paris, 1789. By permission of the Houghton Library, Harvard University.

196. *Dessein en perspective d'une grande loupe . . .*, engraving by Charpentier, Paris, 1782. Photo from the original in the Conservatoire National des Arts et Métiers, Paris.

197. Gesner, Konrad, *De omni rerum fossilium genere*, Zurich, 1565, and *Historia animalium*, Zurich, 1558. By permission of the Houghton Library, Harvard University.

198. By permission of the Houghton Library, Harvard University.

199. Steno, Nicolaus (or Stensen, Niels), *De solido intra solidum naturaliter contento, dissertationis prodromus*, Florence, 1669. From the facsimile reprint, edited by W. Junk, Berlin, 1904.

200. Kircher, Athanasius, *Mundus subterraneous*, Amsterdam, 1665. (The first edition is often said to have been published in 1664 rather than 1665; a later edition appeared in 1678.) By permission of the Houghton Library, Harvard University.

201. Library of the Museum of Comparative Zoology, Harvard University; on deposit at the Houghton Library.

202. Scheuchzer, Johann Jakob, *Museum diluvianum*, Zurich, 1716. By permission of the Houghton Library, Harvard University.

203. Vallisnieri (or Vallisneri), Antonio, *Lezione accademica intorno all' origene delle fontane*, Venice, 1715. By permission of the Houghton Library, Harvard University.

204. By permission of the Houghton Library, Harvard University.

205. Saussure, Horace-Bénédict de, *Voyages dans les Alpes*, Neuchâtel, 1779–1796, plate 3. By permission of the Houghton Library, Harvard University.

206. Photo courtesy of Prof. Alexander M. Ospovat.

207. By permission of the Houghton Library, Harvard University.

208. Faujas de Saint-Fond, Barthélemy, *Histoire naturelle de la montagne de Saint-Pierre de Maestricht*, Paris, 1799. By permission of the Houghton Library, Harvard University.

209. Harvard College Library.

210. Cuvier, Georges, "Notice sur la squélette d'une

très-grande espèce de quadrupède," in *Magasin encyclopèdique*, vol. 1 (1796). Harvard College Library.

211. By permission of the Houghton Library, Harvard University.

212. From the facsimile reprint of the map of Piri Rais (1513) in *Piri Reis haritasi*, edited by Yusuf Akçura, Istanbul (Istanbul Society for Turkish Historical Research), 1935. Harvard College Library Map Collection.

213. Breydenbach, Bernhard von, *Die Heyligen Reysen*, Speier, 1495. Albert E. Lownes Collection, Brown University Library.

214. The British Library, London.

215. The British Library, London.

216. Rumphius (or Rumpf, or Rumph), Georg Eberhard, *Thesaurus imaginum piscium, testaceorum ut et cochlearum, quibus accedunt conchylia*, Leiden, 1711. By permission of the Houghton Library, Harvard University.

217. From the anonymous *Mémoire instructif sur la manière de rassembler, de préparer, de conserver, et d'envoyer les diverses curiosités d'histoire naturelle; auquel on a joint un mémoire intitulé: Avis pour le transport par mer, des arbes . . .*, A Paris et se vend à Lyon, Chez J. M. Bruyser, 1758. Library of the Arnold Arboretum, Harvard University.

218. Frances Loeb Library, Graduate School of Design, Harvard University.

219. Add. MS 15508, f. 11. The British Library, London.

220. Peabody Museum, Harvard University; photo by Hillel Burger.

221. Ashley, Anthony, *The Mariners Mirrour*, London, 1558 (the date is sometimes questioned), translated from Lucas Janssen Wagenaer, *Spieghel der Zeevaerdt*.

222. Waters, David W., *The Art of Navigation in England*, London, 1958. Reproduced through the courtesy of David W. Waters.

223. By permission of the Houghton Library, Harvard University.

224. Picard, Jean, *Mesure de la terre*, Paris, 1671, plate 1. The British Library, London.

225. Copyright Royal Geographical Society, London.

226. National Maritime Museum, Greenwich, England: photo courtesy of Derek Howse.

227. National Maritime Museum, Greenwich, England; on loan from the Ministry of Defence (Navy).

228. MS 2386, fol. 262/3. Trew Collection, Universitäts-Bibliothek, Erlangen.

229. By permission of the Houghton Library, Harvard University.

230. By permission of the Houghton Library, Harvard University.

231. Hales, Stephen, *Vegetable Staticks*, London, 1727, plate 28. The Countway Medical Library, Boston.

232. See note for Illustration 231, plate 8. Library of the Arnold Arboretum, Harvard University.

233. Museum Boerhaave, National Museum of the History of Science and Medicine, Leiden.

234. Library of the Arnold Arboretum, Harvard University.

235. Universitetsbiblioteket, Uppsala, Sweden.

236. Gray Herbarium, Harvard University. See also note for Illustration 242.

237. Sprengel, Christian Konrad, *Das entdeckte Geheimnis der Natur im Bau und in der Februchtung der Blumen*, Berlin 1793. Library of the Arnold Arboretum, Harvard University.

238. Dodart, Denis, *Mémoires pour servir à l'histoire des plantes*, Paris, 1676. Bibliothèque Nationale, Paris.

239. British Museum (Natural History) Botanical Library, Fothergill Album, Tab. 2. Copyright British Museum (Natural History).

240. Engraving by Georg Ehret, Print Room of the Victoria and Albert Museum, London. Photo by John R. Freeman & Co., London.

241. Engraving by Jan Wandelaar, after a painting by Georg Ehret, in Linnaeus's *Hortus Cliffortianus*, Amsterdam, 1737.

242. Engraving (aquatint, stipple, and line) dated 1 December 1799 by J. C. Stadler after a painting by Philip Reinagle, from *The Temple of Flora*, the final part of (or supplement to) Robert John Thornton's *New Illustration of the Sexual System of Linnaeus*. We may agree with Wilfrid Blunt that "the elucidation of the various editions and issues of this work is a tangle which we may well leave bibliographers to unravel"; the best source is G. Dunthorne's *Flower and Fruit Prints of the Eighteenth and Early Nineteenth Century*, London, 1938. One complication arises from the fact that "no two copies of this book are alike" (according to Handasyde Buchanan, in the reproduction of *The Temple of Flora*, edited by Geoffrey Grigson, London, 1951), "and this inevitably means that to some extent the bibliographer's work must be a matter of opinion." *The Temple of Flora* consists of a series of plates "of atlas folio size . . . originally issued in parts between 1798 and 1807" as a supplement to—or as part of—the *New Illustration of the Sexual System of Linnaeus* and then "further published in book form [in 1807] with a contents sheet which lists 31 plates—3 frontispieces and 28 flower plates." Other plates were added, evidently to the as-yet unsold copies, one dated 1 January 1812, and "two plates from the earlier parts of the *New Illustration* frequently make their way into *The Temple of Flora*."

There was also an issue of *The Temple of Flora* in 1810 (1811), plus some portfolios of the plates without text, and a quarto edition of 1812. By permission of the Houghton Library, Harvard University.

243. University of Pennsylvania; photo courtesy of Lynne R. Kressly.

244. Belon, Pierre, *L'histoire de la nature des oyseaux*, Paris, 1555. By permission of the Houghton Library, Harvard University.

245. Rondelet, Guillaume, *L'histoire entière des poissons*, Lyon, 1558. By permission of the Houghton Library, Harvard University.

246. Ruini, Carlo, *Dell' anatomia e dell' infirmità del cavallo*, Bologna, 1598. By permission of the Houghton Library, Harvard University.

247. Casseri, Giulio, *Da vocis auditusque organis historia anatomica*, Ferrara, 1601. The Countway Medical Library, Boston.

248. Fabricius ab Aquapendente, Hieronymus (or Girolamo Fabrizio), *De formatione ovi et pulli*, Padua, 1621. The Countway Medical Library, Boston.

249. Photo courtesy of the Museo di Storia delle Scienze, Florence.

250. Swammerdam, Jan, *Biblia naturae*, Leiden, 1737–1738. (This edition in 3 volumes has Dutch and Latin text in facing pages; the Dutch title reads *Bybel der Natuure*. The Countway Medical Library, Boston.

251. Rijksmuseum voor de Geschiedenis der Natuurwetenschappen, Leiden.

252. See note for Illustration 250.

253. By permission of the Houghton Library, Harvard University.

254. Malpighi, Marcello, *De formatione pulli in ovo*, London, 1673. (F. J. Cole, in *Early Theories of Sexual Generation*, Oxford, 1930, p. 211, indicates that a few copies are dated 1672 "and there is no doubt but that it was first published in that year." Malpighi's *De ovo incubato*, London, 1675, is based on a "Fol. MS dated—October 1672.") Illustration 254 comes from the reprint of *De formatione pulli in ovo* in Malpighi's *Opera omnia*, Leiden, 1687, vol. 2. The Countway Medical Library, Boston.

255. Engraving by William Cowper. Harvard College Library.

256. The Countway Medical Library, Boston.

257. Seba, Albert, *Locupletissimi rerum naturalium thesauri accurata descriptio et iconibus artificiosissimus expressio*, Amsterdam, 4 vols., 1734, 1735, 1758, 1765. The seven-headed Hydra appears in vol. 1. By permission of the Houghton Library, Harvard University.

258. By permission of the Houghton Library, Harvard University.

259. Bibliothèque Centrale, Muséum National d'Histoire Naturelle, Paris.

260. Bibliothèque Nationale, Paris.

261. Harvard College Library.

262. Harvard College Library.

263. By permission of the Houghton Library, Harvard University.

264. University of Pennsylvania; photo courtesy of Lynne R. Kressly.

265. Spallanzani, Lazzaro, *Expériences pour servir à l'histoire de la génération des animaux et des plantes*, Geneva, 1785, a translation (by Jean Senebier) of the sections on generation of Spallanzani's *Fisica animale e vegetabile*, Venice, 1782, 3 vols. (An earlier edition in 2 volumes was entitled *Dissertazioni di fisica animale e vegetabili*, Modena, 1780; the sections on digestion were translated into French by Senebier and published in Geneva in 1785; English versions appeared in London in 1784 and 1789.) Bibliothèque Centrale, Muséum National d'Histoire Naturelle, Paris.

266. This shore bird from Senegal was described and pictured by Louis Bosc in 1792 in *Actes de la Société d'Histoire Naturelle de Paris*. Library of the Museum of Comparative Zoology, Harvard University.

267. Copyright Royal Academy of Arts, London.

268. Burndy Library.

269. Burndy Library.

270. Burndy Library.

271. Burndy Library.

272. Engraving by Andries Jacobzs. Stockius, after a drawing made in 1616 by Jacques de Gheyn; see Paaw, Pieter, *Primitiae anatomicae de humani corporis*, Amsterdam, 1633. Bibliothèque de la Faculté de Mèdecine de Paris, Université de Paris.

273. Engraving by Willem Swanenburgh, after J. C. Woudanus (or van't Woudt), 1616. Academisch Historisch Museum der Rijksuniversiteit, Leiden.

274. Copyright Foundation Johan Maurits van Nassau Mauritshuis, The Hague.

275. The Countway Medical Library, Boston.

276. Pen and ink drawing in the Windsor Royal Library, Courtauld Institute of Art, No. 19058 verso. Reproduced by gracious permission of Her Majesty Queen Elizabeth II.

277. Drawing in the Windsor Royal Library, Courtauld Institute of Art, No. 12610, fol. 4 recto. Reproduced by gracious permission of Her Majesty Queen Elizabeth II.

278. Vesalius, Andreas, *De humani corporis fabrica*, Basel, 1543, Book 2, p. 170. Burndy Library.

279. See note for Illustration 278, Book 2, p. 184.

280. See note for Illustration 278, Book 7, fig. 12.

281. Valverde, Juan de, *Historia de la composición del cuerpo humano*, Rome, 1556. The Countway Medical Library, Boston.

282. (There appears to be some confusion as to whether this edition was published in 1626 or 1627.) The Countway Medical Library, Boston.

283. The Countway Medical Library, Boston.

284. The Countway Medical Library, Boston.

285. MSS Haller, 29, pp. 133, 136. Burgerbibliothek, Bern.

286. See note for Illustration 285.

287. Albinus, Bernhard Siegfried, *Tabulae sceleti et musculorum corporis humani*, Leiden, 1747. The Countway Medical Library, Boston.

288. Gautier D'Agoty, Jacques, *Exposition anatomique maux vénériens sur les parties de l'homme et de la femme*, Paris, 1773. The Countway Medical Library, Boston.

289. Oil painting in the National Portrait Gallery, London. Photo courtesy of the Royal College of Physicians.

290. Haller, Albrecht von, *Deux mémoires sur le mouvement du sang et sur ses effets de la saignée*, Lausanne, 1756. The Countway Medical Library, Boston.

291. Vesalius, Andreas, *Tabulae anatomicae sex: Six Anatomical Drawings* [of] *Andrew* [sic] *Vesalius*, London, privately printed, William Stirling Maxwell, 1874. National Library of Medicine.

292. Drawing in the Windsor Royal Library, Courtauld Institute of Art, Quaderni I, fol. 3-Windsor 19062 recto-S 55,72. Reproduced by gracious permission of Her Majesty Queen Elizabeth II.

293. Harvey, William, *Exercitatio anatomica de motu cordis et sanguinis in animalibus*, Frankfurt, 1728. The Countway Medical Library, Boston.

294. Descartes, René, *Traité de l'homme*, Paris, 1664. The Countway Medical Library, Boston.

295. Malpighi, Marcello, *De pulmonibus observationes anatomicae*, Bologna, 1661. The Countway Medical Library, Boston.

296. Sachs de Lewenheimb (or Sachs von Lewenheimb), Philip Jakob, *Oceanus macro-microcosmicus*, Vratislavia, 1664, frontispiece. By permission of the Houghton Library, Harvard University.

297. Haller, Albrecht von, *Icones anatomicae*, Göttingen, 1743–1754.

298. Mascagni, Paolo, *Vasorum lymphaticorum corporis humani historia et iconographia*, Siena, 1787. The Countway Medical Library, Boston.

299. Vesalius, Andreas, *De humani corporis fabrica*, Basel, 1543, Book 5, p. 378, fig. 25. Burndy Library.

300. Rüff (or Rueff), Jacob, *De conceptu et genera-tione hominis*, Zurich, 1554. By permission of the Houghton Library, Harvard University.

301. Drawing in the Royal Windsor Library, Courtauld Institute of Art, folio 8-Windsor 19102 recto. Reproduced by gracious permission of Her Majesty Queen Elizabeth II.

302. Spieghel, Adriaan van den, *De formatu foetu liber singularis . . .*, Padua, 1626. The Countway Medical Library, Boston.

303. MS E, after p. 123. The Dutch Society of Sciences.

304. *Philosophical Transactions of the Royal Society*, no. 81, 25 March 1672, tab. 2; the article about "Dr. Kerkringius" appears on pp. 4018–4023. By permission of the Houghton Library, Harvard University.

305. From *Anatomia corporum adhuc viventium, qua docet Theophrastus Paracelsus*, in *Aurora Thesaurusque Philosophorum Theophrasti Paracelsi . . .*, Basel, 1577; the figure is reproduced in Pagel, Walter, *Paracelsus*, New York-Basel, 1958.

306. Cohn (or Cohen), Tobias, *Maaseh tuvish*, Venice, 1708. Library of Congress.

307. Descartes, René, *Traité de l'homme*, Paris, 1664. The Countway Medical Library, Boston.

308. Swammerdam, Jan, *Tractatus physico-anatomico medicus de respiratione usuque pulmonum*, Leiden, 1667. The Countway Medical Library, Boston.

309. By permission of the Houghton Library, Harvard University.

310. Reproduced from photographs of original drawings, on glass-plate negatives, at the Conservatoire National des Arts et Metièrs, Paris. Photo courtesy of M. Daumas.

311. Harvard College Library.

312. The Countway Medical Library, Boston.

313. Library of Congress.

314. Blumenbach, Johann Friedrich, *De generis humani varietate nativa liber*, Göttingen, 1776 (perhaps published in 1775, according to the *Dictionary of Scientific Biography*); third edition, Göttingen, 1795, from which this illustration is taken. The Countway Medical Library, Boston.

315. "Negres [d'Afrique]," anonymous engraving. Cabinet des Estampes, Bibliothèque Nationale, Paris.

316. Harvard College Library.

317. Dodart, Denis, *Mémoires pour servir à l'histoire des plantes*, Paris, 1676, frontispiece. By permission of the Houghton Library, Harvard University.

318. Rome, MS Archivio Linceo 4, cc. 251v., 252r.

319. Vascellini, G., *Serie di ritratti d'uomini illustri toscani*, Florence, 1773. By permission of the Houghton Library, Harvard University.

320. The Royal Society, London. Photo by John R. Freeman & Co., London.

321. The Royal Society, London. Photo by John R. Freeman & Co., London.

322. Presentation copy inscribed by the author. (This copy was formerly in the private collection of E. N. da C. Andrade; see Sir Harold Hartley, ed., *The Royal Society*, London, 1960, p. 137.) Photo courtesy of the Royal Society, London.

323. The Royal Society, London. Photo by John R. Freeman & Co., London.

324. The Royal Society, London. Photo by John R. Freeman & Co., London.

325. Bibliothèque Nationale, Paris.

326. Tachard, Gui, *Voyage de Siam des Pères Jésuites*, Paris, 1686. Library of the University of California, Berkeley.

327. I. Bernard Cohen collection.

328. Detail from Illustration 327.

329. This print was reengraved and published in La Condamine, Charles Marie de, *Journal du voyage fait par ordre du roi à l'équator*, Paris, 1751. I. Bernard Cohen collection.

330. Det Kongelige Bibliotek København.

331. M. V. Lomonosov Museum, Leningrad.

332. Ceruto, Benedetto, and Chiocco, Andrea, *Musaeum Franc. calceolari Benedicto Ceruto, medico, incoeptum, et ab Andrea Chiocco descriptum et perfectum*, Verona, 1622. Library of the Museum of Comparative Zoology, Harvard University.

333. Wingendorp, G., *Museum Wormianum, seu historia rariorum tam naturalium quam artificialium tam domesticarum quam exoticarum . . .*, Amsterdam, 1655, frontispiece. Library of the Museum of Comparative Zoology, Harvard University.

334. Vincent, Levinus, *Description abrégée des planches qui représentent les cabinets et quelques unes des curiosités contenues dans le Théatre des merveilles de la nature de Levin Vincent*, Haarlem, 1719. Library of the Museum of Comparative Zoology, Harvard University.

335. Ruysch, Frederik, *Museum anatomicum Ruyschianum*, Amsterdam, 1691. The Countway Medical Library, Boston.

336. Seba, Albert, *Locupletissimi rerum naturalium thesauri accurata descriptio*, Amsterdam, 1758, vol. 3. (This four-volume set was published in 1734, 1735, 1748, and 1765.) By permission of the Houghton Library, Harvard University.

337. Scalberge, Frédéric, *Le Jardin du Roy*, 1636. Bibliothèque Centrale, Muséum National d'Histoire Naturelle, Paris.

338. Bibliothèque Nationale, Paris.

339. Charles Eames photo, Office of Charles and Ray Eames, Venice, Calif.

340. National Library of Medicine.

341. Birmingham Library, England.

342. I. Bernard Cohen collection.

343. Humphrey, H., "The CowPock—or—the Wonderful Effects of the New Inoculation," in *Publications of the Anti-Vaccine Society*, London, 10 June 1802. The Countway Medical Library, Boston.

344. Charles Eames photo, Office of Charles and Ray Eames, Venice, Calif.

345. Bibliothèque, Nationale, Paris.

346. *Le Magnétisme dévoilé*, engraving. Bibliothèque Nationale, Paris.

347. Lent to the Science Museum, London, by the Rt. Hon. the Earl of Cork and Orrery; purchased by the Science Museum in 1974. Photo courtesy of the Science Museum, South Kensington, London.

348. Collection of Historical Scientific Instruments, Harvard University.

349. Derby Museum and Art Gallery, Derby, England.

350. Photo courtesy of the Science Museum, South Kensington, London.

351. I. Bernard Cohen collection.

352. M. V. Lomonosov Museum, Leningrad.

353. M. V. Lomonosov Museum, Leningrad.

354. Engraving by J. B. Tilliard, after J. B. Leprince. I. Bernard Cohen collection.

355. Robichon de la Guérinière, François, *Ecole de cavalérie*, Paris, 1733, part 3, chap. 1, headpiece. Philip Hofer collection, Cambridge, Mass.

356. Harvard College Library.

357. Nollet, Jean-Antoine, *Leçons de physique expérimentale*, Paris, 1749, vol. 5. Harvard College Library.

358. See note for Illustration 357.

359. Rosenbach Foundation, Philadelphia.

360. Barlet, Annibal, *Le Vray et méthodique cours de physique resolutive*, Paris, 1653. By permission of the Houghton Library, Harvard University.

361. The British Library, London.

362. I. Bernard Cohen collection.

363. Harvard College Library.

364. Harvard College Library.

365. Harvard College Library.

366. Harvard College Library.

367. Harvard College Library.

368. By permission of the Houghton Library, Harvard University.

Index

The numbered references in this index refer either to textual material, which is designated by page numbers in lightface, or caption material, which is designated by illustration numbers in **boldface**. It is suggested that, in addition to the page and illustration number references given here for any particular subject, the reader also consult the Picture Sources and Credits section for any illustration listed, which may provide some additional information.